Hands-On Chemical Ecology

Dietland Müller-Schwarze

Hands-On Chemical Ecology

Simple Field and Laboratory Exercises

 Springer

Dietland Müller-Schwarze
College of Environmental Science
   and Forestry
State University of
   New York-Syracuse
1 Forestry Drive
Syracuse NY 13210
USA

ISBN 978-1-4419-0377-8   e-ISBN 978-1-4419-0378-5
DOI 10.1007/978-1-4419-0378-5
Springer Dordrecht Heidelberg London New York

Library of Congress Control Number: 2009926240

Cover insert photograph and all photographs in the text by the author

Printed on acid-free paper

Springer is part of Springer Science+Business Media (www.springer.com)

Preface

Chemical ecology is still a young science with innumerable "growing points", covering organisms from microbes to humans, and levels from molecular to eco-systems. In the near future more and more colleges are likely to offer courses and practical exercises in this field. The author hopes that once a collection of field and lab exercises is available, it will help accelerate this trend. This is the first book of practical exercises in chemical ecology.

Large numbers of students are growing up in urban areas, with limited access to animals and plants in their natural environments. The field experiments sketched out in this book serve at least three purposes: They provide to many students a close look at, if not an intricate knowledge of, the natural history of particular plants and animals and their habitat for the first time, and secondly, they teach insights into otherwise invisible behavioral and ecological processes such as chemical communication or interactions between predator and prey. Both are basic to the third purpose, the intel-lectual processes of planning experiments, gathering data and arriving at fact-based conclusions. Far from the hubbub of city and campus and the many other competing activities, the quiet of the fields, forests and lakes allows total immersion into the web of relationships among organisms and focusing on the scientific question at hand.

This book grew out of two decades of teaching chemical ecology courses such as "Introduction to Chemical Ecology", "Chemical Ecology of Vertebrates", and "Laboratory and Field Exercises in Chemical Ecology" at the College of Environmental Science and Forestry of the State University of New York in Syracuse. The first is an undergraduate course for biology and chemistry students. The latter two were graduate courses, with advanced undergraduates also admitted. During the last several years the author has developed exercises for the laboratory and for the field. The exercises are simple and do not require sophisticated equip-ment. They can be carried out at colleges as well as in high schools. This collection of experiments focuses on the ecological aspect of chemical ecology; isolation and identification of compounds that mediate ecological effects are outside the purview of the book.

It is hoped that college and university undergraduate and graduate students will find the exercises useful. Biology teachers in high schools might also engage their students in these experiments and thus introduce them to the fascinating world of "ecology with a chemical bend".

Acknowledgments. The author thanks all those who over the years have contributed to the courses, particularly the students whose enthusiasm alone was extremely rewarding. Special thanks go to Maryann Page (née Schwoyer) whose dedication as Teaching Assistant in two courses was unsurpassed, and who developed the tannin radial diffusion assay for us. Matt Petit provided great support in a subsequent course. I thank Dr. David Jones, University of Florida, for the cyanide assay and the supplies he generously donated to us. The project benefited from help by colleagues at our SUNY ESF campus. Special thanks to my long-time colleague Dr. Stephen Stehman for statistical advice. Dr. Kimberly L. Schulz contributed chapter 15. Dr. José Giner on our campus helped with advice and literature on chemical aspects, and Dr. Stephen Teale contributed the moth pheromone experiment in chapter 25. Drs. Timothy D. Paine, Fred M. Stephen and Melissa Fierke contributed the exercise on resin defense by conifers. All photos are by the author.

Syracuse, NY   Dietland Müller-Schwarze

Contents

Section I   Field Experiments

**1 Mutualism: Effects of Ants on Aphids, Their Predators,
and Host Plants** 3

    Materials Needed.. 4
    Procedure.. 5
    Results 5
    References 6

2 Predatory Fish Responses to Prey Odors (Chemical Lures)................ 7

    Materials Needed.. 8
    Procedure.. 8
    Results 11
    References 11

**3 Sour Grapes: Methyl Anthranilate as Feeding
Repellent for Birds**.. 13

    Materials Needed.. 14
    Procedure.. 15
    Results 16
    References 17

4 Responses of Small Mammals to Predator Odors in the Field 19

    Materials Needed.. 22
    Procedures 22
    Results 23
    References 24

5   **Squirrels' Avoidance of Predator Odors** 25

    Introduction 26
    Materials Needed 27
    Procedure 27
    Results 29
    References 29

6   **Squirrels, Acorns, and Tannins** 31

    Procedures 33
    Results 34
    References 34

7   **Field Grid for Testing Winter Feeding by Rabbits or Cottontails** 37

    Materials Needed 38
    Procedure 38
    Results 40
    Reference 42

8   **Food Choices by Mammalian Herbivores and the Role
    of Defense Compounds: Example Beaver** 43

    Procedure for Part 1 46
    Procedures for Experiments 2–5 46
    References 50

9   **Scent Marking in Free-Ranging Mammals. Examples:
    Beaver or Badger** 51

    Materials Needed 53
    Procedures 53
    References 57

10 **Capsaicin as Feeding Repellent for Mammals** 59

    Materials Needed 60
    Procedure 60
    Results 61
    References 61

11 **Search for "Chemical Ecology Stories" in the Forest
    or Other Ecosystem** 63

    Northeastern Forest in North America 64
    Amazon Forest 65
    References 65

Section II   Laboratory Experiments

12   Test for Cyanogenic Compounds in Plants 69

Introduction 70
Materials 71
Method 71
Results 72
References 73

**13   Herbivory and a Simple Field Test for Total
Phenolics in Trees** 75

Materials Needed 77
Method 77
References 80

14   Radial Diffusion Assay for Tannins 81

Materials Needed 82
Procedure 82
Results 83
Previous Results 84
References 84

15   Chemically Induced Defenses in Phytoplankton 85

Introduction 85
Materials Needed 86
Procedure 87
Results 89
References 90

**16   Induced Defense: Herbivory on Juvenile vs. Adult
Growth Stages of Trees** 91

Materials Needed 92
Procedure 92
References 96

**17   Jasmonic Acid Effect on Plant Volatiles (or How
to Make a Fern Smell Like a Rose)** 97

Materials Needed 97
Procedure 97
Results 98
References 98

18 Effect of Tannins on Insect Feeding Behavior 101

    Materials Needed 102
    Procedure 102
    Results.. 103
    References.. 104

**19 Leaf Disk Test: Bioassay of Effect of Tannins on
    Insect Feeding Behavior** 105

    Materials Needed 106
    Procedure 106
    Results.. 107
    References.. 108

20 Two-Way Choice Test for Social Odors in Mice 109

    Materials Needed 111
    Procedure 111
    Results.. 111
    References.. 112

21 Scent Marking in Mice: Open Field Test 115

    Materials Needed 116
    Test 1a: Sex Differences 117
    Test 1b: Effect of Age of Urine.. 117
    Test 1c: Scent Marking by Different Types of Males 117
    Test 1d: Responses to Urines from Different Types of Males 118
    Test 2: Effect of Absence of Scent Marks................................... 118
    Results.. 118
    Conclusions.. 119
    Some Previous Results... 119
    References.. 119

22 Human Body Odor Discrimination: T-Shirt Experiment 121

    Materials Needed 122
    Procedure 122
    Results.. 123
    References.. 124

23 Coping with Plant Volatiles in Spicy Food ("Burping Exercise") 125

    Procedure 126
    Results.. 127
    Some Previous Results... 127
    References.. 128

Section III Perspectives of Experiments: Development and Application

24 Miscellaneous Experiments Chosen by Students................................ 131

25 Further Possible Experiments... 133

Intraspecific Communication: Pheromones 133
Interspecific Responses: Allomones (Emission of Chemicals
that Benefit the Sender)... 139
Interspecific Stimuli: Kairomones (Compounds Used
for the Benefit of the Receiver); Responses to Prey Chemicals................. 141
Interspecific Stimuli: Kairomones; Prey Responses
to Predator Chemicals 144
Allelopathy.. 146
Environmental Odors 147

Index... 149

Introduction

The exercises collected here represent a microcosm of the vast range of potential chemical ecology projects. Some require relatively little time, such as Chaps. 12 (test for cyanogenic glycosides), 17 (jasmonic acid effect) and 23 (burping). Others, especially field experiments such as Chaps. 4 (small rodent responses to predator odors), 7 (cottontail winter feeding) or 8 (food choice of free-living rodents), tend to be more timeconsuming. They are better suited for courses at field stations where large blocks of time on two or more consecutive days are available.

While by virtue of geography the animals and plants for the described experiments are northeastern North American species, equivalent other species of mammals, birds, insects, or plants in other areas of the world will be equally (or even better?) suited.

The author has strived to use simple language, with a minimum of jargon. The introductions to the exercises are held short in the assumption that lectures on the topics have preceded the practical part, or that the instructor will provide some background at the start of the laboratory session. Instructors can be, and will have to be, very flexible in the choice of exercises, organisms, settings, and time allotted. Therefore, the manual is not intended to be a rigid, "canned" set of instructions, but rather a "pick-and-choose" proposition, with opportunities to elaborate into many directions. Some exercises such as those dealing with the bird repellent (Chaps. 3), the cyanide test (Chaps. 12), and herbivory on different plant growth types (Chaps. 15), work consistently well with clear-cut results virtually every time, while some field experiments are more open-ended, particularly given their relative complexity and the time constraints a formal college course entails. Ambiguous results make for great discussions. The experiments should be cleared with the Institution Animal Core Commitee.

The exercises cover the range from practising certain frequently used research tools such as the grid experiment (Chaps. 7), leaf disk test (Chaps. 19), T-maze (Chaps. 20), and Open Field (Chaps. 21), to asking questions and designing tests to answer them. Several approaches are equally valid. Most students are familiar with *hypothesis testing*. In the *inductive approach* in the tradition of Karl von Frisch, Konrad Lorenz, and Niko von Tinbergen, all three Nobel laureates, a behavior is studied by intensive observation. This leads to the "why" questions. These in turn spawn experiments to elucidate the immediate function, the stimuli governing the behavior, the development of the behavior and the ultimate adaptive value. My teacher

Lorenz emphasized the importance of *observation without preconceived notions* ("voraussetzungloses Beobachten"). Obviously, such observations precede hypothesis testing. Experience, discipline, and ample time are required to go the steps from unstructured observing to asking questions, to finding experimental ways of teasing out from the animal some valid answers. This way, new phenomena are discovered and we can avoid the bandwagon in a rut. Therefore, open-ended exploration, such as Chaps. 11, occupies an important place in field studies.

The author welcomes feedback from readers and users of this book (dmullers@esf.edu). This first collection of exercises in chemical ecology fills a gap, and is at the same time as limited as one would expect from any first version.

Section I
Field Experiments

Chapter 1
Mutualism: Effects of Ants on Aphids, Their Predators, and Host Plants

Aphids, their predators and ant guards. **a**) Elderberry aphids (*Aphis sambuci*) on elderberry. A predatory lady beetle approaches the colony. **b**) Lady beetle larvae feeding on aphids (most likely mealy plum aphid, *Hyalopterus pruni*) on a plum leaf. **c**) A hoverfly larva (marked by arrow) eating a path through a colony of elderberry aphids. **d**) Two ants guard the elderberry aphids. **e**) Ants feeding on honeydew producing elderberry aphids. **f**) Oleander aphids (*Aphis nerii*) on common milkweed (*Asclepias syriaca*). Note the absence of ants. Oleander aphids take up toxic cardenolides from the milkweed (Rothchild & Reichstein 1970), and ants avoid them. This aphid species is aposematically colored (Malcolm 1986). (By contrast, ants tend the much less toxic, greenish species *Aphis asclepiadis*, also on milkweed)

D. Müller-Schwarze, *Hands-On Chemical Ecology: Simple Field and Laboratory Exercises,*
DOI 10.1007/978-1-4419-0378-5_1, © Springer Science+Business Media, LLC 2009

Aphids extract sap from plant stems, specifically the phloem tissue. They excrete "honeydew" which still contains plant sugars. Ants collect this honeydew, often by "milking" the aphids, and use it as food. In return, they protect the aphids against predators. Ants prey on predators of aphids such as ladybird beetles (coccinellids), thus defending the aphids. Ants also shelter aphids by taking them or their eggs into their nests during inclement seasons. In a sense, ants herd aphids like cows. Furthermore, without removal of honeydew, aphid colonies become fouled.

Some aphid species depend on ants (they are *myrmecophilous*) while others do not (*nonmyrmecophilous* aphids). The latter are better at defending themselves: They move faster and defend themselves chemically. When attacked by ladybird beetles, they release the alarm pheromone *(E)*-β-farnesene. In response to the pheromone other aphids walk around or drop from the plant. Pea aphids even may grow wings in response to alarm pheromone, allowing them to fly from the host plant. Ants also prey on other herbivorous insects, thus lowering their impact on the host plant.

The ants discriminate their attended aphids from unattended ones by odor. They will respond less to odor of aphids outside their attended group of aphids. Nonattended aphids will even be attacked and removed by the ants. If unattended aphids are treated with extract from attended aphids, they will be less attacked than unattended ones, but more often than attended aphids (Glinwood et al. 1999).

Relationships between among species of aphids, and between aphids and ants, can grow complex. Common milkweed, *Asclepias syriaca,* hosts three species of aphids: The oleander aphid (*Aphis nerii*), *A. aslepiadis,* and *Myzocallis asclepiadis.* Of these, only *A. asclepiadis* is mutualistic with ants, while presence of ants reduces the per capita growth of the other two species. Therefore, in this system, ants can act as mutualists as well as antagonists (Smith et al. 2008). *A. nerii* is laden with cardenolides from the milkweed (Rothchild et al. 1970), and its orange or yellow color is aposematic, a "warning color" to predators (Malcolm 1986).

The purpose of this experiment is to examine the role of ants in the life of aphids. Although chemical stimuli play an important role in their interactions, here we focus on the results of these interspecific behaviors. Nevertheless, the observer can try to gently touch an aphid with a pair of fine tweezers and observe the responses of ants and other aphids over a distance, presumably in response to an alarm odor.

Materials Needed

1. Glue
2. Thread
3. Ladybird beetles, obtained commercially or collected in the field
4. Data sheet

Procedure

1. Baseline observations:
   (a) Find aphid colonies. Look on plants such as elderberry (*Sambucus* sp.), goldenrod (*Solidago*), steeplebush (*Spiraea*), or wild raisin (*Viburnum* sp.)
   (b) Count number of aphids per stem
   (c) Record ant traffic on stems (tally number of ants seen passing per 5 min)
2. Field experiment:
   (a) Exclude ants from their aphid colonies and study the effects of this manipulation
   - Apply glue to the basal part of five stems. This will prevent ants from visiting their aphid colonies farther out on the plant shoot.
   - Leave five other aphid colonies untreated, as controls. Mark control twigs with colored thread because they have no other mark to recognize them by.
   - Check the number of aphids at the ten stems daily for 3 days.
   - Look for evidence of leaf damage by other herbivorous insects: count leaf holes, notches at leaf edges.
   (b) Study effects of ants on aphid predators:
   - Watch for natural aphid predators such as ladybird beetles or housefly larvae on each stem under observation. Record kind and numbers.
   - Place a lady beetle on one of the ten stems under study. Record behavior of beetle, aphids, and ants. Repeat for each of the remaining nine stems.
   - Record the results in your Data Sheet 1.1.

Results

- Compare numbers of aphids at end of the 3 days (or more, if possible) at treated twigs
- Test significance of difference with a two-sample *t* test (Data Sheet 1.2)
- Do same for control twigs
- Compare plant appearance scores of treated twigs over 3 days

Data Sheet 1.1 Aphid protection by ants

Date:			Time:		Site:			
Plant species	Plant #	Twig #	# aphids	# ants	Behavior	Other insects	Plant appearance score	Remarks

Data Sheet 1.2 Data arrangement for two-sample t test

Twig	Numbers of aphids left after 3 days	
Twig #	Ants excluded	Control
1		
2		
3		
4		
5		
Total		
Mean		
SE		

- Do same for controls
- Test the difference with a two-sample *t* test
- Graph the data

References

Bowers WS, Nishino C, Montgomery ME, Nault LR, Nielson MW (1977) Sesquiterpene progenitor, germacrene A: An alarm pheromone in aphids. Science 196:680–681

Breton ML, Addicott JF (1992) Density-dependent mutualism in an aphid-ant interaction. Ecology 73:2175–2180

De Joy P (1993) Communication between ants (Hymenoptera: Formicidae) and aphids (Homoptera: Aphididae). deepblue.lib.umich.edu/handle/2027.42/54423

Glinwood R, Willekens J, Pettersson J (1999) Discrimination of aphid mutualists by the ant *Lasius niger*: Evidence for odour marking. In: Poster, 16th Annual Meeting International Society of Chemical Ecology, Marseille, France, 13–17 Nov 1999

Malcolm SB (1986) Aposematism in a soft-bodied insect: A case for kin selection. Behav Ecol Sociobiol 18:387–393

Nault LR, Montgomery ME, Bowers WS (1976) Ant-aphid association: Role of aphid alarm pheromone. Science 192:1349–1350

Oliver TH, Jones I, Cook JM, Leather SR (2008) Avoidance responses of an aphidophagous ladybird, *Adalia bipunctata*, to aphid-tending ants. Ecol Entomol 33:523–528

Rothschild M, von Euw J, Reichstein T (1970) Cardiac glycosides in the oleander aphid, *Aphis nerii*. J Insect Physiol 16: 1141–1145

Seibert TF (1992) Mutualistic interactions of the aphid *Lachnus allegheniensis* (Homoptera: Aphididae) and its tending ant *Formica obscuripes* (Hymenoptera: Formicidae). Ann Entomol Soc Am 85:173–178

Smith RA, Mooney KA, Agrawal AA (2008) Coexistence of three specialist aphids on common milkweed, *Asclepias syriaca*. Ecology 89:2187–2196

Stadler B, Dixon AFG (2005) Ecology and evolution of aphid-ant interactions. Annu Rev Ecol Evol Systemat 36:345–372

Chapter 2
Predatory Fish Responses to Prey Odors (Chemical Lures)

Setting minnow traps baited with chemical lure for attracting predatory fish (top), and emptying trap into plastic bag for identification of fish caught overnight

D. Müller-Schwarze, *Hands-On Chemical Ecology: Simple Field and Laboratory Exercises,*
DOI 10.1007/978-1-4419-0378-5_2, © Springer Science+Business Media, LLC 2009

Predatory fish locate their prey primarily by scent or vision. Here, we are concerned with the chemical sense. Chemical hunting is particularly adaptive for nocturnal species or those living in turbid waters. Many marine and freshwater fishes hunt by smell. The chemical compounds responsible for this attraction have been identified. Most of them are amino acids, and particularly active as mixtures of several amino acids.

Chemical lures impregnated with prey scent have been developed for different species of carnivorous fish. Lures for different game fish species are supposed to contain different chemicals, although usually not revealed on the labels of the products. The artificial lures are made of cellulose ether, a polyalkylene glycol, plasticizers, and other chemicals, and are impregnated with amino acids.

In this exercise, we test the efficacy of chemical fish lures in catching small fish in streams and lakes near the campus. These species are not necessarily "sport fish," but any carnivorous species occurring in three different freshwater habitats.

Materials Needed

1. 27 Minnow traps (or fewer)
2. Six plastic bags (1 gallon)
3. Scented and unscented bait in the shape of worms
4. Some wire to attach bait to trap
5. String to tie trap to sticks on shore, preferably color coded
6. Data sheets

Procedure

Select Experimental Sites

Walk along a *stream*; the shore of a *lake*; and around a *pond*. Locate shallow spots where fish traps can be placed. They should be protected from strong stream currents, and wave action in the lake.

Experimental Design

Prepare one set of traps for each of the three locations: Pond, stream, and lake (or whatever bodies of water are available in your area). Each set has the same number of traps. The number in each set depends on the number of treatments. We will use three treatments: Bait without scent, scented bait, and no bait. Therefore, you will need three traps in each set. Prepare three sets for each location. This adds up to

$3 \times 3 = 9$ traps per body of water. The total number of traps needed for all three locations will be 27. Of course, the number of sets and the number of sites can be varied as needed.

This design has the benefit of showing you the diversity of fish in your area.

Alternatively, you can place all traps along a stream or all traps at the edge of a lake. In this case you have three or six replicates for the same treatment and the same habitat.

Bait Traps

Bait each of the baited trap with one artificial worm, according to treatment scented or unscented. Attach bait in the center of the trap so that fish have to enter the trap, and cannot reach bait from outside. Use wire or string to suspend bait in the trap.

Placing Traps

Put baited and control traps in water in clusters of three, one for each treatment. If you use two traps per treatment, make a second cluster of all treatments at each site. This constitutes a *replicate*. Make sure that scented traps are downstream from unscented ones, so that contamination will not interfere with the experiment. Traps should be totally immersed. Choose spots protected from waves or torrential currents.

Checking Traps

Visit your traps once per day. This can be done anytime during the day. Most convenient for the investigator will be the middle of the day. At each check, empty each trap's content into a plastic bag. Identify fish. Record species and numbers of fish of each species on Data Sheet 2.1. Release fish into water. Return trap to its place in water.

Rebaiting

The scented bait is designed to work only in one fishing episode. Therefore it will most likely not be effective over several days. We will replace the bait after 1, maximally 2 days. You may or may not note a decrease in your catch on the second day with the same bait.

Data Sheet 2.1 Field data for fish attractant experiment

Date	Time	Place	Trap #	Treatment	Fish species caught	# of fish caught	Remarks
			1				
			2				
			3				
			4				
			5				
			6				
			7				
			8				
			9				
			10				
			11				
			12				
			13				
			14				
			15				
			16				
			17				
			18				
			19				
			20				

Data Sheet 2.2 Number of fish species

Site and trap cluster #	Bait w/o scent	Scented bait	Control	Total
Pond 1				
Pond 2				
Pond 3				
Total				
Mean/SE				
Stream 1				
Stream 2				
Stream 3				
Total				
Mean/SE				
Lake 1				
Lake 2				
Lake 3				
Total				
Mean/SE				

Data Sheet 2.3 Total number of fish caught (all species combined)

Site	Control	Unscented bait	Scented bait	Total
Pond				
Stream				
Lake				
Total				
Mean/SE				

Results

The main question is whether the scented lure attracts more fish than the unscented one. Therefore, we organize the data according to treatment, as shown in Data Sheet 2.2:

- Draw bar graphs from these data
- Write report.
- Emphasize your conclusion on the efficacy of the scented bait.
- Also, discuss different fish communities at the different locations.
- How specific are the scent lures? How do fish species differ in their responses? (In our experience, some species such as brown bullhead entered all traps, regardless of whether they were scented or not. Others, such as rock bass and yellow perch, were caught only in scented traps, but were attracted to both trout-scented and pumpkinseed-scented traps equally.)
- To test whether the differences between the treatments are significant, you can use Tukey's test (Data Sheet 2.3)

References

Jones KA (1991) A case for taste. In-Fisherman. Book 101 (June/July/Aug) 31–44

Nuhfer AJ, Alexander GR (1992) Hooking mortality of trophy sized wild brook trout caught with artificial lures. North Am J Fish Manag 12:634–644

Schisler GJ, Bergersen EP (1996) Postrelease hooking mortality of rainbow trout caught on scented artificial baits. North Am J Fish Manag 16:570–5782 Predatory Fish Responses to Prey Odors (Chemical Lures)

Chapter 3
Sour Grapes: Methyl Anthranilate as Feeding Repellent for Birds

Bird repellent experiment: Location of four feeders, each containing birdseed treated with a different concentration of methyl anthranilate

D. Müller-Schwarze, *Hands-On Chemical Ecology: Simple Field and Laboratory Exercises,*   13
DOI 10.1007/ 978-1-4419-0378-5_3, © Springer Science+Business Media, LLC 2009

Birds such as starlings attack Concord grapes less than other grape varieties. Concord grapes contain comparatively high levels of methyl anthranilate (MA):

Methyl anthranilate
(2-aminobenzoic acid methyl ester)

This compound smells like commercial grape juice. A series of studies confirmed that MA (and related compounds such as dimethyl anthranilate), when applied to seeds or grass, inhibits birds from feeding. Methyl anthranilate repels a variety of birds, such as red-winged blackbirds, starlings, pigeons, jungle fowl, herring gulls, ring-necked pheasants, and Canada geese (Avery and Decker 1994; Avery et al. 1995; Cummings et al. 1995; Glahn et al. 1989; Marples and Roper 1997; Mason et al. 1989). This repellent may play an important role in protecting feed grain on farms from bird predation. It is available under the trade names ReJeX-iT, Bird Shield, and Avigon. In this context, it is important to know that MA does not deter mammals. For a review of bird repellen see Spurr (2007).

In this exercise, we test the repellent effect of MA with free-ranging local birds. We asked the questions:

1. How general is the effect on birds? Specifically, to what extent do several species of wild woodland birds - as opposed to those who visit our backyards – avoid MA-treated food?
2. Do different species differ in the strength of their responses to MA?
3. Is the effect concentration dependent?

Materials Needed

1. Methylanthranilate (100 g)
2. Ethyl alcohol (1 l)
3. Shelled sunflower seeds (5 or 10 kg). Used mixed bird seed to attract more bird species.
4. Pans or trays (4) for treating and drying sunflower seeds (capacity about 2 l)
5. A scale (50–1,000 g range)
6. Permanent marker (2), black
7. Bird feeders (5 or 10), cylindrical, 800–1,000 ml capacity
8. Binoculars
9. Notebook (the paper kind)

Procedure

Preparation of Feeders

Calibrate each bird feeder: A 800-ml feeder will hold about 500 g of shelled sunflower seeds. Weigh 50 g, fill into feeder, and mark level with permanent marker. Add another 50 g, mark, and continue until the entire feeder is marked at 50-g intervals. It now looks like (and is) a graduated cylinder.

Search for good places to hang the feeders.

Prebaiting

Fill one or two feeders with untreated sunflower seeds and hang them at the study site. The experiment can begin as soon as the birds have found the food.

Treatment of Seeds

Place four trays (in a pinch: aluminum foil) on lab tables. Weigh out four 500-g portions of sunflower seeds (without shells), one for each tray.

We use three concentrations of MA: 0.5%, 1%, and 2% by weight of the seed. Ethyl alcohol serves as vehicle for MA: for the 0.5% concentration, dissolve 2.3 ml (2.524 g) MA in 100-ml EtOH and pour over the 500 g seeds. Mix. For the 1% MA concentration, use 4.6-ml (5.05 g) MA in 100-ml EtOH, and 8.9-ml (10.1 g) MA in 100-ml EtOH for the 2% treatment. Treat the fourth batch of seeds with EtOH only, as control. Let soaked seeds in trays dry overnight to evaporate the solvent. Use a fume hood or other well-ventilated area to minimize the MA odor in the room.

Fill the feeders. The fifth feeder will be filled with untreated sunflower seeds.

For a double-blind test, a person other than the field observers codes the feeders with letters A–E (or numbers) and safeguards the code.

Placement of Feeders

Find places to hang the feeder. Use hooks that facilitate easy changing of feeder location. Hang the feeders in random order where birds are active. Locate them apart enough (several meters) so that you can clearly see under which feeder ground feeding birds are foraging. But hang them close enough so you can observe all of them at the same time, and they all experience the same conditions in terms of vegetation, sunlight, human, and pet traffic, among others. Rotate positions of feeders twice a day, to counteract any location effects.

Observation of Birds

Prepare a data sheet that contains seven columns for date, time, bird species, number of birds, sex, number of feeder visited, and activity and other remarks (Feed, perch only, chase other bird).

Select a vantage point. Use a bench or chair to sit on. Choose a distance that permits clear identification of bird species and bird behavior at the feeders. Observe for 20 min at a time. Observe for eight such observation periods, spanning 4 days. In summer, mornings (between 6:30 and 8:30) and afternoons (between 17:00 and 18:30) are good times (there is less bird traffic in mid-day).

After each observation period, i.e., in midday and evening, record the level of seeds in each feeder, giving you eight data points for each treatment.

Results

- Tabulate and graph consumption for each treatment, as determined by reading levels of seeds in feeders.
- Tabulate and graph numbers of birds at various feeders, day by day, separate lines for treatments.
- Do simple statistical tests. Average food consumption for the 20-min periods on each day. Use Friedman's two-way analysis of variance for randomized blocks (the days are the blocks, constituting one factor, the treatments are the second factor); followed by pairwise comparisons. Compare two treatments of particular interest, such as lowest MA concentration vs. EtOH only.
- *Write report:* Present results. Discuss findings, especially concentration effects, but also species differences, and any ecological considerations with regard to location of feeders and bird activity (Data Sheets 3.1–3.3).

Data Sheet 3.1 Levels (ml) of remaining sunflower seeds in the various feeders (for recording in field)

Date	Time	Feeder 1	Feeder 2	Feeder 3	Feeder 4	Bird species seen

Data Sheet 3.2 Amount of sunflower seeds removed: Difference between full feeder and level of seeds now (Derived from Sheet 1)

Feeder 1	Feeder 2	Feeder 3	Feeder 4

Data Sheet 3.3 Chemical bird repellent: Behavior of birds at feeder

Date	Time	Species	Number	Sex	Behavior	Remarks

References

Avery ML, Decker DG (1994) Responses of captive fish crows to eggs treated with chemical repellents. J Wildl Manage 58:261–266

Avery ML, Decker DG, Humphrey JS, Aronov E, Linscombe SD, Way MO (1995) Methyl anthranilate as a rice seed treatment to deter birds. J. Wildl Manage 59:50–52

Cummings JL, Pochop PA, Davis JE, Krupa HW (1995) Evaluation of Rejex-it AG-36 as a Canada goose grazing repellent. J Wildl Manage 59:47–50.

Glahn JF, Mason JR, Woods DR (1989) Dimethyl anthranilate as a bird repellent in livestock feed. Wildl Soc Bull 17:313–320

Marples NM, Roper TJ (1997) Response of domestic chicks to methyl anthranilate odour. Anim Behav 53:1263–1270

Mason JR, Adams MA, Clark L (1989) Anthranilate repellency to starlings: Chemical correlates and sensory perception. J Wildl Manage 53:55–64

Spurr EB (2007) Bird control chemicals. In: Pimentel D (ed) Encyclopedia of pest management, vol II. CRC, Boca Raton, FL, p 523

Chapter 4
Responses of Small Mammals to Predator Odors in the Field

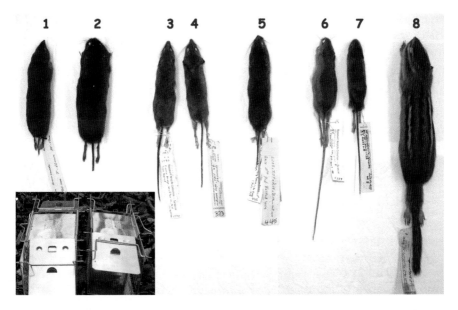

Responses of small mammals to predator odors in live traps. *Top row*: Mammal species that can be expected to enter live traps in and near a northeastern hardwood forest. (1) Short-tailed shrew (*Blarina brevicauda*). (2) Meadow vole (*Microtus pennsylvanicus*). (3) Deer mouse (*Peromyscus maniculatus*). (4) White-footed mouse (*Peromyscus leucopus*). (5) Southern red-backed vole (*Clethrionomys gapperi*). (6) Woodland jumping mouse (*Napaeozapus insignis*). (7) Meadow jumping mouse (*Zapus hudsonicus*). (8) Eastern chipmunk (*Tamias striatus*). A reference collection like this one should be taken into the field for species identification. Inset: Scented live traps can be placed in pairs, with the open entrances facing in opposite directions

D. Müller-Schwarze, *Hands-On Chemical Ecology: Simple Field and Laboratory Exercises*,   19
DOI 10.1007/978-1-4419-0378-5_4, © Springer Science+Business Media, LLC 2009

While we are asleep, forests, meadows, and fields teem with nocturnal mammals. Dramatic scenes of predation take place, all unseen by humans. In this exercise, we will learn how to indirectly record such nocturnal interactions between predator and prey.

Mammalian predators possess keen senses of smell and hearing, ensuring their meals. The prey animals, on the other hand, are adept at detecting and avoiding predators. Again, olfaction plays an important role in this survival mechanism.

Although prey species must coexist with predators, they reduce predation risk by monitoring whereabouts and activity of their main predators and avoiding them to the greatest extent possible in both space and time. Most small mammals rely on olfaction to detect and avoid predators. They can smell mammalian predators such as wolves, coyotes, foxes, cats, or mustelids directly, or extract information about time and place of their activities from predator sign such as droppings, urine, scent marks with secretions from skin glands, tracks, rubs, or scrapings. Rodents can distinguish the odors of different carnivore species. Herbivorous prey species tend to respond more strongly to the odors of sympatric predators than to those of allopatric carnivores (Müller-Schwarze 1972).

For each field experiment with a particular small mammal species we need to know the particular predator(s) preying on, or even specializing on, this prey species in our specific geographical region. To examine how specific the response of the prey may be, we can present odors of different kinds of predators.

There is strong evidence that sulfur compounds in urine and feces signal "predator" to prey species (Nolte et al. 1994). The major compounds in mustelid anal gland that repel small rodents are sulfur compounds such as sulfides, thiols, thietanes, or dithiolanes. These compounds have been used to deter rodent pests (e.g. Sullivan et al. 1988).

Since most small rodents are nocturnal, their responses to predator odors cannot easily be observed directly. Instead, indirect recording techniques are needed. The classical method of choice is catching small mammals overnight in live traps scented with predator odors. Examples are avoidance responses of rodents (and also shrews) to red fox odor in traps (Dickman and Doncaster 1984) and *Apodemus* sp. and *Microtus* sp. to stoat scent in traps (Gorman 1984). The live-trap technique demonstrated that negative responses to predator odors can be deeply ingrained: Orkney voles avoid red fox odor, although there have been no foxes or other carnivores (except for otters, *Lutra lutra*, that do not prey on voles) since the voles were introduced on the Orkney Islands by Neolithic settlers about 5,500 years ago. Such responses are "ghosts of predation past" (Calder and Gorman 1991). Another technique is to apply predator odors on or near food and record overnight feeding responses. This has been done with beavers, *Castor canadensis* (Engelhart and Müller-Schwarze 1995; Rosell and Czech 2000).

In the northeastern woodlands of North America coyotes (and originally wolves), gray and red foxes, mustelids, and nowadays feral dogs and cats are the major predatory mammals. (Other important predators such as great horned owls, or hawks, are not considered here, for obvious reasons). Ground-dwelling small

Data Sheet 4.1 Field record of small mammals caught in traps scented with predator odor

Date:	Time:	Weather:		
Trap #	Odor	Species	Sex	Remarks

mammalian prey species range from chipmunks to woodland jumping mice and short-tailed shrews (see caption of Figure).

We will examine whether the small mammals vulnerable to predation will avoid fresh predator odors in their home ranges. For this purpose we place in the woods live traps that are scented with odors from red fox and wolf. This experiment teaches not only basic ecology, but also the very practical skill how to test potential chemical repellents for rodent pests.

This first of two experiments on predator odor avoidance deals with nocturnal burrowing mammals. The second will involve day-active mammals, such as squirrels (see Chap. 5).

Materials Needed

1. 100 small mammal live traps (for rodents, including chipmunks, and shrews)
2. Predator droppings from a zoo (about 300 g)
3. Cow manure (about 300 g)
4. 500-ml Ethyl alcohol
5. Filter paper (circular, 25–30 cm diameter; and sheets for lining traps)
6. Eyedropper
7. Oatmeal or sunflower seeds as bait
8. Cotton for nesting material
9. Latex gloves

Procedures

Prepare Extracts

Weigh fox and/or wolf droppings. Pour 300-ml ethyl alcohol over 300 g droppings. (We use EtOH because methyl alcohol can be toxic if inhaled or ingested.) Stir. Filter the slurry through a funnel with a large filter paper. Do the same with herbivore (cow) dung. This will yield between 200- and 250-ml extract. Divide each of the two extracts into at least two bottles for safety, in case one gets spilled in the field.

Select Study Site

Choose an area that is uniform in terms of vegetation, soil, exposure, and disturbances such as human or pet traffic. Trap in one homogeneous wooded area and another open one such as a clearing in the woods. Each test area should be at least 30 × 30 m large.

Scenting of Traps

Wear latex gloves to prevent human scent from interfering with the experiment. Cut filter paper so it fits on the bottom of live trap. Place into the trap. With an eyedropper, apply 0.12 ml of scent solution on the filter paper.

Bait the trap with one-half teaspoon of oatmeal or ten sunflower seeds. Also place a wad of cotton into trap. The animals can use this for a token nest to avoid hypothermia while above ground during cool and/or wet nights.

Use four treatments: predator odor, herbivore odor, solvent (alcohol) only, and untreated traps. Each treatment will be replicated with 25 traps. (If two predator odors are being used, 20 traps for each of the five treatments).

Placing Traps

Design a grid. If 100 traps are available, use 50 for the wooded area and 50 for the clearing. Lay out a grid of 5 × 10 traps. If the terrain permits, space the traps 10 m apart. Draw numbers from a hat to place traps of each treatment in a random pattern.

Trapping Routine

Set traps in evening, in summer between 18:00 and 20:00 h. Check early the following morning: between 05:30 and 07:15. Rodents (and particularly shrews) confined to above-ground traps will easily die of exposure and hunger. Identify each trapped animal by species, and also sex and age class if possible and release it where it was caught. Spring all traps. Replace filter paper and cotton so that scent of previously caught animals does not interfere with later captures. Bait every trap with sunflower seeds every night. Repeat for 5 days.

Results

- Enter data in sheet.
- Since numbers for each species will be small, lump all species into one category "small mammals."
- Enter totals in Data Sheet 4.2
- Graph data in a histogram

Data Sheet 4.2 Compilation of data: Numbers of small mammals caught

Rodents caught	Treatment				
	Controls				
	Predator odor	Herbivore Odor	Solvent only	Blank	Total
No					
Yes					
Total # traps					

- Test for statistical differences, particularly between captures in predator-scented traps and controls. Combine numbers for blanks and herbivore odor. Arrange data in Data Sheet 4.2. Use a 2×2 chi-square (χ^2) test.
- Discuss species differences of response.
- How do species differ between the wooded and the open area?
- What do repeat captures tell us about the behavior of the animals?

References

Calder CJ, Gorman ML (1991) The effects of red fox *Vulpes vulpes* faecal odours on the feeding behaviour of Orkney voles *Microtus arvalis*. J Zool Lond 224:599–606

Dickman CR, Doncaster CP, (1984) Responses of small mammals to Red fox (*Vulpes vulpes*) odour. J Zool Lond 204:521–531

Engelhart A, Müller-Schwarze D (1995) Responses of beaver (*Castor canadensis*) to predator chemicals. J Chem Ecol 21:1349–1364

Gorman ML (1984) The response of prey to Stoat (*Mustela erminea*) scent. J Zool Lond 202:419–423

Jedrzejewski W, Jedrzejewska B (1990) Effect of a predator's visit on the spatial distribution of bank voles: Experiments with weasels. Can J Zool 68:660–666

Müller-Schwarze D (1972) Responses of young black-tailed deer to predator odors. J Mammalogy 53:393–394

Nolte DL, Mason JR, Epple G, Aronov E, Campbell DL (1994) Why are predator urines aversive to prey? J Chem Ecol 20:1505–1516

Sullivan TP, Crump DR, Sullivan DS (1988) Use of predator odors as repellents to reduce feeding damage by herbivores. III. Montane and meadow voles (*Microtus montanus* and *Microtus pennsylvanicus*). J Chem Ecol 14:363–377 Results Responses of Small Mammals to Predator Odors in the Field4   Responses of Small Mammals to Predator Odors in the Field

Chapter 5
Squirrels' Avoidance of Predator Odors

Repellent test for squirrels: (**a**) setup of grid of acorns near trees that provide refuge for the squirrels; (**b**) squirrels approaches for another acorn; and (**c**) consumption of an acorn. Note how repeated visits by squirrels have not disturbed the arrangement of the acorns. (**d**) Circular arrangement of acorns

D. Müller-Schwarze, *Hands-On Chemical Ecology: Simple Field and Laboratory Exercises,*
DOI 10.1007/978-1-4419-0378-5_5, © Springer Science+Business Media, LLC 2009

Introduction

This second experiment with predator odors deals with day-active mammals whose behavior can be observed directly and readily. Small mammals such as squirrels are prey to many predatory birds and mammals. Vigilance vis-à-vis predators encompass all major senses: smell, vision, and hearing. In the chemical sphere, predators leave signals from scent marks, droppings, and urine in the environment. Squirrels as typical rodents have a keen sense of smell capable of detecting such predator odors and extracting information such as how recent the "sign" is.

We will examine whether a diurnal rodent, the gray squirrel (*Sciurus carolinensis*) avoids predator odors and whether this avoidance is specific to certain predator species that pose more of a threat than others. This experiment arose from our course "Chemical Ecology of Vertebrates" during the autumn of 2000. Surprisingly, we could not find published studies of predator odor effects on squirrels. Dr. Frank Rosell, then a student in the course, undertook this experiment as his individual research project and extended it after the end of the course for a publication (Rosell 2001).

Unlike many other mammals, squirrels are active during the daytime. We can observe their choices directly, or simply by the results of their actions, here the food choices they made. In North America, the ubiquitous squirrels and chipmunks offer themselves for behavioral experiments in backyards, city parks, cemeteries, National and State parks, and on college campuses. There the animals are conditioned to humans so that experiments can be carried out without disturbing their behavior.

Arboreal mammals, like squirrels, face a greater danger from climbing predators such as mustelids or raccoons than from ground predators such as foxes, coyotes, or wolves. Accordingly, we can compare the responses of squirrels to these two types of carnivores.

Although the squirrels' main predators are owls, red foxes (*Vulpes vulpes*) also prey on gray squirrels when given an opportunity. Humans have a complex relationship with gray squirrels. Many of us are only too aware that they raid bird feeders, gnaw on buildings and tree barks, and nest in attics. In rural areas they also feed on corn. Therefore, squirrels are often persecuted. Hunters pursue gray squirrels to varying extent in different regions of the USA.

The predator scents typically used in experiments are urine, extracts of feces, scent gland products, or combinations of these. Behavioral responses of small mammal to predator odor stimuli range from vigilance to avoiding the site, and feeding inhibition. We can test squirrels' responses to odors of an arboreal predator (cat), a ground predator (fox), and to humans (in most areas harmless pedestrians, but in others they are squirrel hunters), and compare them with their behavior toward odors of a nondangerous herbivore, such as deer or cattle.

Squirrels accept many types of food bait. These include acorns, hickory nuts, butternuts (*Juglans cinerea*), walnuts, peanuts, sunflower seeds, corn (maize), and more. Predator odors can be placed near these foods, or applied to them.

Materials Needed

1. *Food*, such as acorns, bitternuts, walnuts, or peanuts.
2. *Predator odors*, from a zoo, from domestic animals or a commercial source such as a supplier of hunting and trapping lures. Possible choices: *Arboreal* predators such as cat or raccoon; and *ground* predators such as fox or human.
3. Control odor: *herbivore* urine (from cow, goat, sheep, or deer)
4. It is easiest to purchase predator scents from a local outdoor sports store or a mail order business were they are sold as lures to hunters and trappers. However, the exact composition (ingredients and their amounts) of the scent lures is not known in most cases. To be sure the scent is fresh and from one species only and from a specified sex and age class, we recommend to obtain the material from a local zoo and extract and concentrate it in one's own laboratory. The zoo personnel are usually very helpful. They may also allow the experimenter to scoop up the material from a pen him/herself.

Procedure

Stimulus Preparation

The instructors have already obtained predator urine from a commercial supplier or the local zoo. They are from red fox *(V. vulpes)* and an arboreal predator such as the house cat.

Field Site Selection

The food bait, combined with the predator odor, is placed in an area frequented by squirrels such as a park, cemetery, campground, picnic area, or front or backyard of a school or private residence. The food can be placed next to the scent, or treated directly with it. In the author's experience, gray squirrels visit a feeding station more in the morning than afternoon. Find an area where squirrels are active. Place food near trees from which the animals can descend and retreat to in case of alarm. Prebait the squirrels with food to be used in the experiment. In our study, we will use acorns of red oak.

Scent Application

Place a filter paper (size: about 12 cm diameter) on level ground where squirrels are expected to forage. With an eyedropper or disposable pipette, apply 5 ml of scent extract to the filter paper. Place five red oak acorns around the center, weighing down the filter paper. Arrange the acorns or nuts in a regular pattern such as rows, squares, or circles so that missing items can easily be accounted for. If needed, pin down filter paper with a nail. Wear plastic or rubber gloves to avoid contamination with human scent. For a replicate, use two such predator-scent treated groups of five acorns. Place two more groups of acorns on filter paper treated with herbivore urine. Finally, place five acorns each on two untreated, dry filter papers as blanks. All groups of acorns should be equally spaced. Place them near a tree or other vertical object that squirrels like to use as a lookout.

Three- or two-sample choices work best. A predator odor can be juxtaposed to a nonpredator odor and a nonscented control, making for a three-way choice test. In a second version of the experiment, one can differentiate the predator odor by comparing odors of a ground and an arboreal predator. Finally, odors of a native (sympatric) predator can be compared with those of an exotic (allopatric) one (Müller-Schwarze 1972). These test whether sulfur compounds common to many carnivores are alarming – as suggested by Nolte et al. (1994) – or whether more specific stimuli are at work.

Try to observe the behavior of the squirrels directly: After placing the scented food, observe for 30 or 60 min from a distance of about 30 m.

- Note the behavior of the squirrels: Which item do they approach first? Do they sniff before deciding to take an item or to move on to the next? (Squirrels are remarkably nimble: They select items to eat without disturbing the arrangement of the rest, see Figure). If individual squirrels can be distinguished, the better.
- If direct observation is not possible, leave samples alone and return in 2–3 h to check on results, and again after 6 h, but certainly before dark.
- Repeat observations for 3 days.

Data Sheet 5.1 Feeding responses of gray squirrels to red oak acorns

Date	Time	Treat-ment	Acorn pile #	Responses				
				# removed	# eaten	Shells left	# remaining	Remarks

Results

- Tabulate and graph data for numbers of acorns consumed after the observation periods (Data Sheet 5.1).
- Test for significance among the different treatments, using Friedman's test for related samples. The days are blocks for this test. If significant overall, examine differences between pairs of treatments by the Wilcoxon signed ranks test. Remember, the main question is whether predator odor reduces feeding.

As a variation of the experiment described above, one of our course participants chose to present free-ranging gray squirrels with urine of red fox, raccoon (*Procyon lotor*), humans, and - as control - white-tailed deer (*Odocoileus virginianus*) (Rosell 2001).

References

Müller-Schwarze D (1972) The responses of young black-tailed deer (*Odocoileus hemionus columbianus*) to predator odors. J Mammal 53:393–598

Nolte DL, Mason JR, Epple G, Aronov E, Campbell DL (1994) Why are predator urines aversive to prey? J Chem Ecol 20:1505–1516

Rosell F (2001) Effectiveness of predator odors as gray squirrel repellents. Can J Zool 79:1719–1723

Chapter 6
Squirrels, Acorns, and Tannins

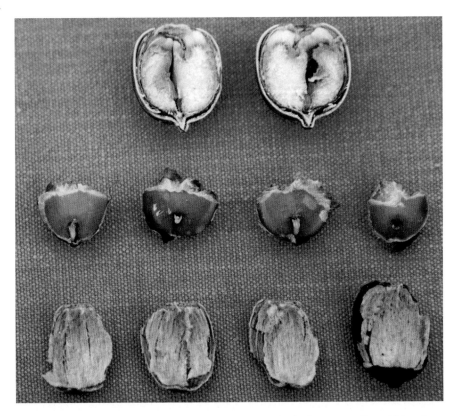

Remains of acorns of red oak, *Quercus rubra*, after gray squirrels, *Sciurus carolinensis*) have been feeding on them. *Top row*: cross section of acorn, with radicle barely visible at apical pole at the bottom. *Middle row*: Parts of acorns left behind by squirrels. They discard apical pole (with radicle visible). Such pieces are later consumed by birds such as blue jays (*Cyanocitta cristata*). *Bottom row*: pieces of acorn shells. Photo: D. Müller-Schwarze

D. Müller-Schwarze, *Hands-On Chemical Ecology: Simple Field and Laboratory Exercises,*   31
DOI 10.1007/978-1-4419-0378-5_6, © Springer Science+Business Media, LLC 2009

Squirrels, but also birds such as jays, bury acorns in the ground to cache them as winter food. By this behavior, they also disperse the acorns and start the germination process by exposing the acorns to soil moisture. Burying protects acorns from surface-feeding competitors such as deer, ruffed grouse, and turkeys. It also prevents other squirrels from pilfering the stores, provided the "owner" remembers where he had buried the seeds, and other squirrels are not attracted by the odor of the buried seeds.

An acorn is technically a fruit, although customarily called a "seed." It consists of a shell (pericarp) that encloses two seedling leaves (cotyledons) which in turn contain food reserves. The cotyledons connect to a tiny seedling by petioles. The seedling is also (technically incorrect) called the "embryo." This embryo is located near the tip (distal end) of the acorn and is folded between the cotyledons. When germination starts, the pericarp splits at the tip. The seedling root (radicle) appears first. Then the entire seedling emerges, as the petioles of the cotyledons arch out downward and stretch to 2–5 cm. The radicle develops into a substantial taproot, while the presumptive leaves (epicotyl) and stem (hypocotyl) will grow little until spring.

Squirrels certainly do know their taxonomy: they identify acorns as either storable or less suited for underground storage. They bury red oak (*Quercus rubra*) acorns as winter caches, but eat immediately those of white oak, *Q. alba* (Smith and Follmer 1972; Smallwood and Peters 1986). White oak germinates already in fall, while red oak germinates later, and therefore "keeps better" in the ground. White oak sends a thickened taproot deep into the ground before winter starts. It thus "buries itself" and is thought to escape seed predation this way (Fox 1982). To prevent white oak acorns from germinating, gray squirrels kill these seeds by excising the seed "embryo" before caching them. Mature squirrels practice this technique more often than juveniles (Fox 1982). The squirrels discard the embryo.

Chemical ecologists have been debating the role of secondary plant compounds in feeding behavior of rodents. Specifically, in this context, do squirrels choose acorns with low tannin content as food? High tannin content can inhibit squirrels from consuming acorns, while high levels of fat attenuate the effects of tannins (Smallwood and Peters 1986). Tannins bind with proteins which by the way, is the basis for tanning of skins, transforming them into leather. An herbivore is adversely affected when tannins bind to proteins in the food, to proteins on the surface of the mouth cavity, and to important enzymes of the animal.

Tannin levels vary within an acorn. They are more concentrated in the apical (pointed) portion with the embryo. Accordingly, gray squirrels, and also grackles and jays, selectively consume more often the basal part of an acorn than the distal part (Steele et al. 1993). The embryo will still germinate after being left over by a squirrel. The chemical gradient in the acorn may represent an adaptation that results in a compromise: animals consume only the part of the acorn that contains the least tannin, but the better protected seedling and parts around it will still survive the predation.

Another question that has spawned several studies is how squirrels and other seed-burying animals find their treasure again. Experiments have shown that squirrels use memory and landmarks, and perhaps their sense of smell, and among birds,

Clark's nuthatches (Nucifraga columbiana) rely primarily on memory and land-marks (Vander Wall 1982). Deermice, for instance, find buried oats, wheat, and barley grains better if they are scented with safflower oil (Howard et al. 1968).

A particularly intriguing third question is: does burying of acorns lower their high tannin content to render them more palatable? In other words, do squirrels practice food processing? So far, the few experiments conducted have yielded inconclusive results. Chung-MacCoubray (1993) and Dixon et al. (1997) did not find significantly changed tannin levels after experimental storage in soil.

We will try to test ourselves whether free-ranging (or captive) squirrels discriminate between acorns that have been buried for several weeks and those that have not. In a later laboratory experiment we will analyze tannin levels in the apical and basal poles of acorns, and also compare buried acorns with untreated controls in this regard. We will use the Radial Diffusion Assay for Tannins (Chap. 14).

Instead of burying acorns in the ground, we can soak them in water in the laboratory. Native Americans used to dry, shell, and then soak acorns in net bags in water, often with lye (NaOH and/or KOH, obtained by soaking wood ash in water) to remove the high levels of tannins (Moerman 1998). Even today, acorns are being used for cooking and baking. The tannins are removed by either boiling the shelled acorns in water, or by soaking coarsely ground acorns in cold water. In either case, the process takes hours or days, with several or many changes of water.

Procedures

Preparation of Experiment

1. Red oak acorns were buried in fall when the acorn crop fell to the forest floor and squirrels buried acorns. To protect the acorns from predation during their time in the ground, they were enclosed in a hardware cloth cage. This has been done by the instructors. The acorns were 15 cm under the surface, with a layer of oak leaves on top which in turn was covered by 20 cm snow. After 3 months, the acorns were unearthed again and are available now for your experiment. (Obviously, this experiment is best suited for a winter or spring semester.)

2. Prebaiting: Immediately before the experiment, determine whether the local squirrel population is in a burying or feeding mode, or both: Prebait gray squirrels near the campus with counted piles of red oak acorns for several days. This way they will expect food at your chosen site and will return to look for food.

3. Weigh 20 acorns each of the conditioned (buried) batch and the untreated control acorns. Does their weight differ? Which ones are heavier? How do you interpret your results?

4. Main experiment: Provide equal numbers of conditioned and control acorns to free-ranging, but prebaited gray squirrels. You will be provided with 60 acorns of each of the two types. Run three replications of 20 treated and 20 control acorns each. Observe for 20 min after placing the 20 plus 20 acorns:

Data Sheet 6.1 Squirrels' selection of acorns conditioned in soil or water

Date	Time	Acorn pile #	Responses				
			# Acorns removed	# Acorns eaten	# Shells left	# Acorns remaining	Remarks

Results

- How do the squirrels approach the acorns?
- How do they seem to choose? By smell, touch, or taste?
- What do they do with the acorns? Distinguish the following behaviors: Handle, mouth, gnaw, eat, divide, excise embryo, carry away, bury? Any other behavior?
- In what sequence do they deal with the acorns? One type first?
- If the squirrels did not respond during the 20 min, return 30 min later to see the results of their activity. This way you have missed most of the behavior, but you can record the number of acorns of each type consumed or removed.
- Reduce the data; test for significance with a 2×2 χ^2 test (Treated/untreated vs. eaten or not eaten).
- Graph the results (Data Sheet 6.1).

References

Chung-MacCoubray AL (1993) Effects of tannins on protein digestibility and detoxification activity in gray squirrels (*Sciurus carlinensis*). MSc Thesis VPA & State University, Blaxcksburg, VA

Dixon MD, Johnson WC, Adkisson CS (1997) Effects of caching on acorn tannin levels and blue jay dietary performance. Condor 99:756–764

Fox JF (1982) Adaptation of gray squirrel behavior to autumn germination by white oak acorns. Evolution 36:800–809

Howard WE, Marsh RE, Cole RE (1968) Food detection by deer mice using olfactory rather than visual cues. Anim Behav 16:13–17

Koenig WD (1991) The effects of tannins and lipids on digestion of acorns by acorn woodpeckers. Auk 108:79–88

Moerman DE (1998) Native American ethnobotany. Timber Press, Portland, OR

Müller-Schwarze D, Brashear H, Kinnel R, Hintz KA, Lioubomirov A, Skibo C (2001) Food processing by animals: Do beavers leach tree bark to improve palatability? J Chem Ecol 27:1011–1028

Schmidt KA, Brown JS, Morgan RA (1998) Plant defenses as complementary resources: a test with squirrels. Oikos 81:130–142

Smallwood PD, Peters WD (1986). Grey squirrel food preferences: The effects of tannin and fat concentration. Ecology 67:168–174

Steele MA, Knowles T, Bridle K, Simms EL (1993) Tannins and partial consumption of acorns: Implications for dispersal of oaks by seed predators. Am Midl Nat 130:229–238

Tomback DF (1980) How nutcrackers find their seed stores. Condor 82:10–19

Vander Wall S (1982) An experimental analysis of cache recovery in Clark's nutcracker. Anim Behav 30:84–946   Squirrels, Acorns, and Tannins

Chapter 7
Field Grid for Testing Winter Feeding by Rabbits or Cottontails

Field grid for testing cottontail or rabbit feeding on twigs treated with repellent. *Top*: Freshly set up grid. *Bottom*: Remains after several nights of feeding by cottontails. Note that some rows are completely demolished, while much is left in row on *far right*

D. Müller-Schwarze, *Hands-On Chemical Ecology: Simple Field and Laboratory Exercises,*
DOI 10.1007/ 978-1-4419-0378-5_7, © Springer Science+Business Media, LLC 2009

This field experiment requires 4–6 days. Setup on day 1 requires about 2 h, and daily checking of the results on the following days requires about 30–60 min.

Food choice experiments with free-ranging animals in the field have many advantages over tests in the laboratory or with fenced-in subjects, because they happen in the "real world." Any practical applications of repellents or attractants will eventually occur in this real world, regardless how they have been tested before. However, the field presents a significant disadvantage to the experimenter: During the growing season free-ranging animals enjoy a vast food base against which any bait placed by the experimenter is infinitesimal. In winter, by comparison, the food supply of herbivores has shrunk to a few survival foods. Therefore, in winter mammals accept the bait much more readily, and experiments can yield very good results in terms of food discrimination, provided the animals are not extremely starved.

In a typical experiment, twigs of palatable plants such as apple trees are coated with a repellent or feeding inhibitor. This renders the food less attractive. We can address different questions:

How do different compounds, mixtures of compounds, or complex plant extracts compare in their repellent effects?

How does concentration affect the repellent effect of one particular compound?

Do extracts from different plant parts (leaves, flowers, stem, and roots) differ in their repellent effects?

Do herbivores become less selective, i.e., accept more repellent, over the course of the winter?

Materials Needed

1. Twigs of apple, pear, or cherry trees. Already existing prunings from orchards are perfect for our purpose: 100 twigs for a 10 × 10 grid.
2. The repellent: a commercial herbivore repellent, a plant extract, or known aversive plant compounds.
3. Solvent: Methyl alcohol or ethyl acetate
4. A large graduated cylinder (100 ml).
5. A pair of snippers.
6. Rubber gloves.
7. Blender.
8. A ruler or yardstick (ca. 88. c)

Procedure

Stimulus Preparation

Any herbivore repellent can be tested. Good choices are commercially available deer or rabbit repellents.

You can also prepare your own plant extract, as in the reference below. Use a plant species that herbivores avoid such as spurges (*Euphorbia* spp.) or false hellebore (*Veratrum viride*). Separate plant parts such as leaves, stems, and roots. Grind plant parts separately with ethyl acetate in a blender. (In our work, we extracted 60 g of roots of *Euphorbia lathyris* with 200-ml ethyl acetate. For about 300 g plant tops we used 400-ml ethyl acetate.) Filter the extract, discard the plant tissue.

Finally, you can test any compound or mixture of compounds of interest to you. This may even lead to new discoveries. For this exercise, we assume you will apply five different treatments, including repellents to be tested, a solvent control, and untreated twigs. With 100 twigs in the array, there will be 20 for each treatment. Of course, this number can be varied to suit particular circumstances.

The test solutions will be labeled in code, to insure unbiased data recording in the field.

Field Site

Find a place where cottontails, *Sylvilagus floridanus* (in other regions rabbits, *Oryctolagus cuniculus*) are consistently active.

Find or prepare an open area for placing the food grid. Be sure the site experiences a minimum of disturbance by humans or pets.

Setting up Grid

Cut twigs to same length (about 50 cm). In the field, dip twigs in the test solution in a tall graduated cylinder. This allows you to measure the uptake, averaged over 10 or 20 twigs. Stick twigs 10 cm into ground, 20 cm apart in rows of 10, the row 30 cm apart. One treatment per row works best. A number of neighboring twigs with the same treatment represent a "pseudobush," since herbivores tend to avoid the neighbor(s) of unpalatable shoots, in nature usually part of the same bush. (In a truly random arrangement, palatable neighbors of less palatable items might be avoided, due to this "neighbor effect" or "overshadowing.")

Checking for Results

After nocturnal feeding by the free-ranging animals, record the status of each twig on the next day. At this stage, the samples are coded; the field investigator does not know the treatment for each twig. Use a data sheet with a line for each twig. The variables will be numbers of tips bitten off, length of each twig remaining, and amount of bark peeled (in centimeters, measured with a ruler).

Continue test for 3 or 5 nights of feeding. The longer time may be needed if the animals do not accept the bait easily during the first 1 or 2 nights.

Conversely, you may have to terminate the experiment earlier, if your subjects are voracious and deplete the food supply very fast. Eventually, all palatable twigs (controls) will have their bark eaten, and will be cut into pieces. Twigs with effective repellents will still be wholly or partly intact. You will also note that over the days, as palatable samples will be depleted, the animals will become more likely to also consume unpalatable items. Therefore, it is important to evaluate data day by day.

The experiment can be run with daily replacements of consumed twigs. This complicates the experiment somewhat, as each day an unpredictable number of twigs have to be cut, dipped, and set into the ground.

Results

- Test significance with an analysis of variance for five treatments.
- Display results in a bar graph (Data Sheet 7.1).

Data Sheet 7.1 Results of field grid for food choices by cottontails (or rabbits)

| Chemicals tested: | | | Date: | |
| Site: | | | Remarks: | |
Sample (twig)	Tip removed	Length remaining (cm)	Amount of bark removed (cm)	Remarks
A1				
A2				
A3				
A4				
A5				
A6				
A7				
A8				
A9				
A10				
B1				
B2				
B3				
B4				
B5				
B6				
B7				
B8				
B9				
B10				
C1				

(continued)

Data Sheet 7.1 (continued)

Sample (twig)	Tip removed	Length remaining (cm)	Amount of bark removed (cm)	Remarks
Chemicals tested:			Date:	
Site:			Remarks:	
C2				
C3				
C4				
C5				
C6				
C7				
C8				
C9				
C10				
D1				
D2				
D3				
D4				
D5				
D6				
D7				
D8				
D9				
D10				
E1				
E2				
E3				
E4				
E5				
E6				
E7				
E8				
E9				
E10				
F1				
F2				
F3				
F4				
F5				
F6				
F7				
F8				
F9				

(continued)

Data Sheet 7.1 (continued)

Chemicals tested:			Date:	
Site:			Remarks:	
Sample (twig)	Tip removed	Length remaining (cm)	Amount of bark removed (cm)	Remarks
F10				
G1				
G2				
G3				
G4				
G5				
G6				
G7				
G8				
G9				
G10				
H1				
H2				
H3				
H4				
H5				
H6				
H7				
H8				
H9				
H10				
I1				
I2				
I3				
I4				
I5				
I6				
I7				
I8				
I9				
I10				
Totals				
Mean/SD				

Reference

Müller-Schwarze D, Giner J (2005) Cottontails and Gopherweed: Anti-feeding compounds from a spurge. In: Mason RT, LeMaster MP, Müller-Schwarze D (eds) Chemical Signals in Vertebrates, vol 10. Springer, New York, NY7   Field Grid for Testing Winter Feeding by Rabbits or CottontailsResultsReference

Chapter 8
Food Choices by Mammalian Herbivores and the Role of Defense Compounds: Example Beaver

Beaver picking up experimental sticks in a food choice experiment. The animal will transport the sticks through the water to the lodge. There one or more beavers will consume the bark. The peeled sticks will be released into the water and can be found floating or stranded in the dam

D. Müller-Schwarze, *Hands-On Chemical Ecology: Simple Field and Laboratory Exercises,*   43
DOI 10.1007/978-1-4419-0378-5_8, © Springer Science+Business Media, LLC 2009

Herbivores such as deer, bovids, rodents, or marsupials encounter a great diversity of plant secondary metabolites (PSMs). These PSMs greatly affect the food choices these animals make. In turn, herbivores affect plants by stimulating induced defenses as a consequence of browsing. For food choice experiments in the wild we need animals that can be found predictably in certain places at certain times. The beaver (*Castor canadensis* or *C. fiber*) is such a species. Beavers stay year-round near their lodges and readily accept food provisions.

The beaver is a generalist herbivore. Depending on the local vegetation, many plant species can be beaver's food. These include a large variety of trees, shrubs, grasses, forbs, and aquatic plants. Beavers select food according to palatability. Nutrient content and aversive plant chemicals determine palatability. Moreover, after beavers or other herbivores have clipped or cut shrubs and trees, the following season's regrowth will increase its chemical defenses. The new shoots assume the *juvenile* growth form: compared to *adult* type growth they have larger leaves, unbranched growth, and intensified chemical defenses. In many areas you will most likely see young saplings of the juvenile growth form in aspen, cottonwood, willow, and basswood.

When beavers first colonize a site they begin by harvesting the most palatable species, such as aspen (*Populus* spp.) or willow (*Salix* spp.). Over time, they change the vegetation by their selection of trees for food and construction of dams and lodges. Eventually, often only the least preferred conifers are left, and the beavers move to a new area until some of the depleted vegetation has regenerated. Beavers move back in, and the cycle repeats.

In this exercise, we test the food preferences of local beavers. The experiment teaches how to design, run, and evaluate field experiments.

In this cafeteria-style food choice experiment, we will provide beavers with a range of woody plants, determine their choices, and interpret the results in terms of palatability, determined by chemical plant defenses and nutrient content.

The experiment has five parts that can be performed independently:

1. In a nonexperimental survey we assess the beavers' use of naturally occurring trees and shrubs, and compile an Electivity Index to document the animals' preferences among the trees and shrubs on the site.
2. Compare consumption of different tree species by provisioning the beavers, using only mature, unaltered boughs.
3. Test beavers' preferences between adult and juvenile growth forms.
4. Treat palatable food with compounds known from unpalatable trees and check whether this lowers the beavers' consumption.
5. Present beavers with presoaked sticks to determine whether they prefer food that had some PSMs leached out.

If a field study is not feasible, the students can discuss the data in the three graphs at the end of this chapter (Figs. 8.1–8.3). Form two or three groups and formulate conclusions about feeding strategies of beavers. Compare and debate the possibly quite divergent conclusions.

Procedure for Part 1

Delineate the study area: A strip of vegetation on each bank of a stream, up to 30–50 m from the water; or in the case of a lake or large pond, lay transects perpendicular to the water's edge, up to 60 m from the shore. Transects can be about 5-m wide. Count all trees or shrubs there, and all trees or shrubs that have been cut by beavers, mostly stumps. Relate abundance (numbers of trees present) and utilization for each species separately in an *Electivity Index:*

$$Ej = \ln \frac{(rj)(1-pj)}{(pj)(1-rj)}, \qquad (8.1)$$

where r is the number of utilized trees (or shrubs) of a given species and p is the number of available trees of the same species. An E larger than zero represents preference, smaller than zero avoidance. Values near zero mean a species is taken in proportion to its abundance.

Procedures for Experiments 2–5

Collecting Plant Samples

Cut ten twigs of four deciduous tree species, ranging from palatable such as aspen or willow, to least palatable such as red maple, black locust, or witch hazel. The actual species tested will vary from year to year, and with availability. Each twig is 60-cm long and has its leaves on. For one beaver site, collect ten such bows of each species. If you use more than one beaver site, you need multiple sets of ten twigs each.

Instead of leafy twigs, you can present "minilogs" to the beavers. Cut sticks of about 2-cm diameter and 30-cm long. This offers itself during spring and autumn when trees are leafless.

Placing Samples

Stick boughs into ground at water's edge and near a beaver lodge or "feeding bed"; bed where the animals are most likely to visit and detect the food. Place samples in a row parallel to the water, with the boughs 30 cm apart. While random order might be desirable, think for a moment how the beaver forages: It cuts one cane of a shrub or one of the many clustered saplings of one species. If that proves unpalatable, it will avoid the whole shrub or cluster. Therefore, it is biologically better to group the samples of one kind in a kind of "pseudoshrub." (In a random arrangement, the beaver may avoid the neighbor of an unpalatable sample, just because it stands next to it). Therefore, place ten twigs of one species in a row, and continue the same row with ten of another species. In this one row, all samples are equally distant from the water. The beaver, of course, should have equal access to the whole array, without obstacles. This often requires to clear vegetation from the site.

Place samples in the evening, shortly before beavers emerge from their lodge. This way you minimize the risk that other herbivores, such as deer, will eat or sample the food first. (If that is a real possibility, the twigs have to be fenced off with high chicken wire, leaving 20 cm so open at the bottom for the beavers).

Checking Results

Go to the experimental site the next morning. Record which samples:

(a) Are still in place and appear untouched
(b) Have been sampled (partially eaten, bitten into, pulled out of ground and left there)
(c) Are missing and have presumably been consumed as food or building material (check dam and lodge whether you can find these twigs there)
(d) Have been dragged into water (watch these for consumption during following days)

Repeat these observations every morning for 5 days. Beavers are often slow to accept food during the first night.

If a trail camera is available, you can record the individuals and where and when they feed during the night.

Results

- Tabulate data
- Draw a graph: Days on abscissa, and cumulative consumption on ordinate
- Write report

Part 2

Compare the consumption of 3–4 species of trees, with no other variable considered.

Part 3

Study the beavers' responses to juvenile vs. adult growth forms. Species used will depend on availability. Best are aspen (*Populus tremuloides* or *P. grandidentata*) or cottonwood (*Populus deltoides*).

Part 4

Treat twigs with a known PSM and determine the effect of this treatment on acceptance by the herbivores.

Keep boughs of palatable species such as aspen or willow in a solution of a secondary plant metabolite for 2–3 days. Boughs will take up some of the solution. Use gallic acid, a phenolic. Gallic acid occurs abundantly in red maple and reduces feeding in tent caterpillars, for instance. To prepare saturated solution, dissolve 11.5 g gallic acid in 1 liter of water.

Before taking samples to the beaver pond, paint the stems with the gallic acid solution. Now, this phenolic covers outside and (presumably) inside of the twigs. As a control, use also untreated aspen or willow. Use 12 twigs of each of the two treatments.

Part 5

This part deals with the possible effects of leaching out plant secondary compounds from usually less preferred food. Offer beavers sticks 30-cm long (1-cm diameter) of trees known to contain high levels of phenolics such as witch hazel (*Hamamelis virginiana*) or red maple (*Acer rubrum*). Half of the sticks are untreated, and the other half has been soaked in water for 2–3 days. As control, offer a preferred food such as untreated aspen.

- After tabulating the data, draw a graph.
- Test for statistical significance of the differences in feeding. For multiple comparison, use Cochran Q test; for matched two-sample comparisons (Data Sheet 8.1), McNemar test.

Previous Results

Below I present some results from earlier experiments in our courses. These are meant only as benchmarks. Each locale is unique in terms of trees present, or trees preferred. Also, season will affect food choice. The first graph (Fig. 8.1) shows

Data Sheet 8.1 Beaver food choice

Date: Time: Site:

Date	Sample(Treatment)	Position	# Unaltered	# Missing	# In water	# Sampled	Remarks

preferences among six tree species. The second (Fig. 8.2) compares consumption of adult form and juvenile form willow. The final experiment shown here (Fig. 8.3) includes adult and juvenile form aspen among six tree species. Interpret these data in terms of food preference by beavers. Are the three graphs consistent with one another? Compare your own results with these presented here. Discuss the possible reasons for any differences between these and your findings.

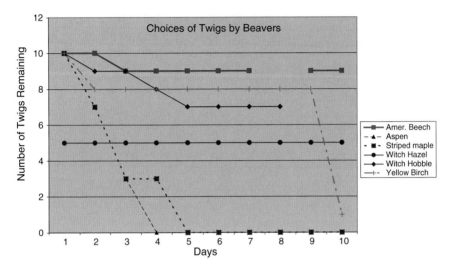

Fig. 8.1 Feeding preferences by free-ranging beavers (*Castor canadensis*) for 3 species of northeastern deciduous trees and shrubs: American beech (*Fagus grandifolia*), quaking aspen (*Populus tremuloides*), striped maple (*Acer pensylvanicum*), witch hazel (*Hamamelis virginiana*), witch hobble (*Viburnum lantanoides*) and yellow birch (*Betula alleghaniensis*). The experiment was run over 10 days. Each morning the number of remaining sticks was counted. Only one of 10 beech sticks was taken after 10 days, while, at the other extreme, no sticks of aspen and striped maple remained after 4 and 5 days, respectively

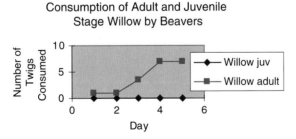

Fig. 8.2 Consumption of adult and juvenile type willow twigs by free-ranging beavers. Vertical axis shows numbers of twigs consumed each day. The experiment was run for 5 days

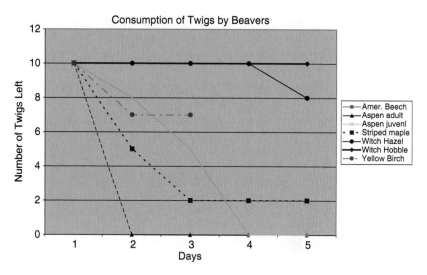

Fig. 8.3 A mixed feeding experiment with free-ranging beavers, including six species of deciduous trees and shrubs, and two growth forms (adult and juvenile) of quaking aspen. The experiment was run for 5 days. Note the two extremes: witch hazel vs. adult form of aspen

References

Abu-Zaid MM, Grant GG, Helson BV, Benninger CW, DeGroot P (2000) Phenolics from deciduous leaves and coniferous needles as sources of novel agents for lepidopteran forest pests. In: Shahidi F, Ho C-T (eds) Phytochemicals and phytopharmaceuticals. AOCS, Champaign, IL, pp 398–416

Basey JM, Jenkins SH, Miller GC (1990) Food selection by beavers in relation to inducible defenses of *Populus tremuloides*. Oikos 59:57–62

Müller-Schwarze D, Schulte BA, Sun L, Müller-Schwarze C, Müller-Schwarze A (1994) Red maple (*Acer rubrum*) inhibits feeding by beaver (*Castor canadensis*). J Chem Ecol 20:2021–2034

Müller-Schwarze D, Brashear H, Kinnel R, Hintz KA, Lioubomorov A, Skibo C (2001) Food processing by animals: do beavers leach tree bark to improve palatability? J Chem Ecol 27:1011–1026

Chapter 9
Scent Marking in Free-Ranging Mammals. Examples: Beaver or Badger

Left: an unusually large scent mound built by beavers. *Right:* Application of a scent sample to the cork on an experimental scent mound

D. Müller-Schwarze, *Hands-On Chemical Ecology: Simple Field and Laboratory Exercises,*   51
DOI 10.1007/ 978-1-4419-0378-5_9, © Springer Science+Business Media, LLC 2009

Beaver scent marking. Top: Beaver approaches, then sniffs the experimental scent mound. Middle: Beaver scratches experimental scent mound with forepaws, then straddles and marks it. Bottom: Beaver leaves the scent mound and swims away

Scent marking plays a central role in the social organization and mating behavior of mammals. This exercise on scent marking can start as a general ecological project that addresses many aspects of behavior and ecology of a particular mammal species in the wild. Once the overall social organization of the species and the particular population has been sketched out, we can narrow the project down to scent communication as essential part of the fabric of an animal's daily life. We can study dogs on campus, cats in our neighborhood, badgers in agricultural areas, or beavers in wooded areas, depending on where we live. Badgers, for example, defend a territory, maintain latrine sites, and mark with their subcaudal gland. The scent marking follows a seasonal pattern, and latrine site use is also correlated with food abundance (Pigozzi 1990).

In the following we focus on scent marking in beavers. Since this species happens to be primarily nocturnal, we study the results of marking, rather than the behavior itself. Many of the questions can also be asked for badgers, for instance. This exercise is particularly suited for field courses in or near wooded areas where beavers occur.

This exercise consists of three parts that can be done in sequence or separately, depending on the time available: General survey of the site, which raises a variety of broader biological and ecological questions; specific study of existing scent marks and their distribution; placing experimental scent marks of a particular description and observing the behavioral responses and products of overmarking.

Materials Needed

For Parts 1 and 2:
1. Binoculars

For Part 3:

1. Beaver castoreum from commercial source, sold as "beaver castor" or "quill"
2. Rubber gloves
3. Gardener's planting trowel
4. Binoculars
5. Several corks (3 cm diameter) or bottle caps

Procedures

Part 1: Survey of an Active Beaver Site

At an active site lives a beaver family, consisting of parents, yearlings, and kits of the year. During the daytime they stay in their lodge which can be a bank lodge or freestanding, surrounded by shallow water. While the animals them-selves are not visible, we see signs of their daily activities, including movements (tracks, trails), feeding (tree stumps, cut off branches and twigs, and peeled sticks), building infrastructure (lodges, dams, and canals), and scent marking (scent mounds).

Procedure

Keep Field Notes on the Following Aspects

1. Where is the site located?
   • At a stream, pond, lake, river, intermittent body of water such as seepage or a ditch?
   • Is this an isolated beaver colony, or are there other ones in the vicinity, i.e., within approximately 3 km? (This will be important for Part 3.)
2. What do you see?
   • If it is a *pond*, describe size; new or old; vegetation around it, aquatic vegetation.
   • Is it good or poor beaver habitat?
   • *Lodge:* Describe size and shape. Located where: Bank or freestanding? Orientation? New or old? Does it appear inhabited? Why? Is it mudded? Are there fresh sticks on top?
   • Other beaver sign: "Transportation system": Are there beaver *trails*? *Slides*? *Canals*? *Tracks* in mud? Do they lead to recent foraging sites?
   • Feeding behavior: Are there tree *stumps*? Are they old or fresh? What tree species? Estimate how recently they were cut. Survey stumps from the water's edge up the slope away from the water. Sample along a 60-m long transect perpendicular to the water's edge. Do tree diameters and species change with distance from the water? Are there cut twigs in the water? If so, what species? Peeled sticks in the water or on the bank? Are twigs and peeled sticks concentrated in a "feeding bed"?
   • Do you see beaver *droppings*? They are best visible in shallow, clear water near the dam. Are they fresh or disintegrated?
   • Sketch a map of the beaver site, including the features mentioned above.

Conclusions

So far, what can we conclude? Is there enough fresh activity that it is clear that the site is inhabited? Can we infer the number of beavers here? Does a single individual, a pair, or a family live here? Are there young offspring, evidenced by small gnawing traces in wood or tracks in mud?

Implications

- Did the beavers invest much in the infrastructure? Is it a site with a good food supply in terms of plant species and abundance? Does this beaver site create habitat for other plant and animal species? Are any rare or endangered plants or animals present? Include any management recommendations.
- If this beaver site were (or actually is) located in a developed area, would it pose any problems, such as flooding or cutting valuable trees? What could be done about it?

Part 2: Survey of Scent Mounds

This experiment is feasible in spring and early summer.

1. Carefully walk around a pond, especially near lodge and dam. Look for *scent mounds*. These are little piles of mud at the water's edge that smell more or less strongly like beaver castoreum, depending on the time since marking.
2. Count the scent mounds. Enter their location in relation to lodge, dam, and trails on a sketched map you made before.
3. How are the scent mounds distributed? Randomly or at "strategic places"?
4. Does the number of scent mounds support the notion of a correlation of marking frequency and population density? Population density here means number of other beaver colonies within a 5 km radius, or within 5 km upstream and downstream.

Part 3: Experimental Scent Mounds

To study beaver responses to defined scent stimuli we place an artificial scent mark near an active beaver lodge and record the animals' responses.

We can observe their behavior directly or check later for any changes at the scent mounds made by the beavers. The first can be difficult because the animals are very sensitive to disturbances such as the presence of humans especially in groups.

Procedure

1. In the afternoon or early evening before the beavers emerge from their lodge, build a scent mound from mud, scooped up from the pond, just as the beavers do. Use a little garden trowel, or in a pinch, a plastic tub. Wear rubber or plastic gloves to avoid contamination with human scent. About 1 l of mud suffices.
2. Place the scent sample on top of the scent mound. If you use commercial, dried, and ground beaver castoreum, sprinkle a teaspoon full over the mud. For dissolved scent samples, place a large cork or bottle cap on scent mound,

    drip 0.25 ml of solution on it. This prevents the sample from running down the
sides of the wet mud mound in an uncontrolled manner.

3. In the evening, before any beavers emerge, sit down downwind of the beavers'
   activity area, on the opposite side of the pond, for instance, if the distance is
   relatively short. Observe, time, and record of the beavers' behavior: approach
   from water onto land, sniffing, pawing with front feet, marking, adding mud to
   the scent mound. If night vision goggles are available, extend observation into
   darkness.

4. On the next day, check for changes at scent mounds: cork removed; mud mound
   disturbed (scratched apart); new mud added; fresh, strong smell; additional
   nearby scent mounds built by beavers overnight.

5. If a wildlife trail camera (camera trap) is available, set it up near the experimental
   scent mounds to record visits by beavers during the night.

Results

- Write a report on your observations.
- Discuss the function of scent marking in the context of the beavers' extensive
  investment in habitat modification.
- This exercise, especially Part 3, focuses on observing behavior. In the time
  available, the observed reactions to the odor stimulus will most likely not be
  frequent enough to lend themselves to statistical analysis (Data Sheet 9.1).

Data Sheet 9.1 Beaver scent marketing

Date	Time	Site	Treatment	Result	Remarks

References

Houlihan PW (1989) Scent mounding by beaver (*Castor canadensis*): functional and semiochemi-
    cal aspects. M.S. thesis, State University of New York College of Environmental Science and
    Forestry, Syracuse, NY.
Müller-Schwarze D, Sun L (2003) The Beaver: Natural History of a Wetlands Engineer, Cornell
    University Press, Ithaca, NY
Pigozzi G (1990) Latrine use and the function of territoriality in the European badger, *Meles
    meles*, in a Mediterranean coastal habitat. Anim Behav 39:1000–1002
Rosell F, Nolet BA (1997) Factors affecting scent-marking behavior in Eurasian beaver (*Castor
    fiber*). J Chem Ecol 23:673–689

Chapter 10
Capsaicin as Feeding Repellent for Mammals

Birds and squirrels at capsaicin-treated food. *Top*: Bluejay (*Cyanocitta cristata*) feeds on capsaicin-treated birdseed, while gray squirrel (*Sciurus cadrolinensis*) simultansously feeds on untreated birdseed. *Bottom*: Simultaneously, bluejay feeds on birdseed treated with capsaicin (a mammal repellent), and a gray squirrel feeds on birdseed treated with methyl anthranilate, a bird repellent

This "real-world" exercise tests the efficacy of a feeding repellent in free-ranging mammals. It constitutes the counterpart to the repellent effect of methyl anthranilate on feeding by birds (see Chap. 3). The experiment works at any place with wild squirrels and in any season. The many mammal repellents on the market are aimed against deer, predators such as raccoons, foxes and coyotes, and rodents such as voles, mice, squirrels, woodchucks, and others. Mammal repellents are known under names such as copper naphthenate, trimethacarb, zinc naphthenate, and ziram.

Capsaicin is an alkaloid and the active flavor component of chili peppers, which belong to the genus *Capsicum*. It is thought to be a deterrent against certain herbivores and fungi. Technically, capsaicin is 8-methyl-*N*-vanillyl-6-nonenamide:

It irritates skin and mucous membranes. In terms of chemoreception, it irritates the trigeminal nerve. Capsaicin binds to the *vanilloid receptor subtype 1* (VR1). Birds lack this receptor. Since they cannot sense capsaicin, they eat and distribute seeds of red hot peppers. Medically capsaicin is used as painkiller (analgesic).

Capsaicin is the active ingredient in commercial squirrel repellents. It is very common that squirrels raid bird feeders. To discourage them from doing that, bird seed can be shaken with powdered capsaicin preparations, so that it adheres to the surface of the seeds. The treated food is supposed to irritate a squirrel's mouth and trigeminal nerve.

The purpose of this exercise is to test whether treatment with capsaicin inhibits feeding by squirrels significantly, and how long this effect lasts, i.e., whether at least some squirrels manage to cope with capsaicin.

Materials Needed

1. 2–4 cylindrical bird feeders
2. Bird seed. Most kinds are attractive to squirrels.
3. Commercial squirrel repellent (or other substances to be tested)

Procedure

Prebaiting: Hang feeder with untreated seed in an area frequented by squirrels. Keep feeding until squirrels discover the food.

After squirrels start visiting the feeder, start the experiment. Use two cylindrical bird feeders. Fill one with untreated bird seed. This is the control.

For the other feeder, treat bird seed with capsaicin: Pour powder over seeds and shake in a closed container. The amount needed will be given on the commercial product. One supplier recommends one rounded tablespoon of powder for 6 cups of bird seed, in metric terms, 8.6 ml powder to approx. 0.5 l seed.

Instead of bird seed, other foods attractive to squirrels such as acorns or peanuts can be treated and offered along with untreated controls. The acorns can be placed in several clusters of 5, equidistant from one another. Alternate with clusters of untreated acorns (or peanuts). Observe squirrels from a distance for 20 min. Record the first response of individual animals to the capsaicin-treated food. Does their behavior betray signs of discomfort? If no squirrels approach the food during that time, return after 1 h, and again after 2 h and tally the numbers of items removed.

Results

- Tabulate the amounts of birdseed (or numbers of acorns, etc.) consumed.
- Compare the two treatments with a two-sample t test.
- How long does the repellent effect last? Do the squirrels manage to cope with the repellent by circumventing it in some fashion, or by habituating to it? To address this question, treat another batch of bait with repellent and offer it on day 2, and a third batch on day 3.
- Do the squirrels find a way to consume the food despite the aversive additive?
- Do individual animals employ different strategies to deal with the treated seed? Since in most places the animals are not tagged, natural marks will help to identify individuals. Some have distinctly notched ears, or patches of hair missing, while others have characteristic fur color patterns.

References

Mason JR, Bean NJ, Shah PS, Clark L (1991) Taxon-specific differences in responsiveness to capsaicin and several analogues: correlates between chemical structure and behavioral aversiveness. J Chem Ecol 17:2539–2552

Wikipedia (as of Nov. 2008) Capsaicin. http://en.wikipedia.org/wiki/Capsaicin#Mechanism_of_action.10

Chapter 11
Search for "Chemical Ecology Stories" in the Forest or Other Ecosystem

Tangarana ants on tangarana tree ("torture tree"; *Triplaris* sp.). Instead of its own plant metabolites, this tree depends on ants to defend it and rewards them with nectar from extrafloral nectaries. Manu National Park, Peru, 15 Dec 2004

D. Müller-Schwarze, *Hands-On Chemical Ecology: Simple Field and Laboratory Exercises,*   63
DOI 10.1007/978-1-4419-0378-5_11, © Springer Science+Business Media, LLC 2009

Northeastern Forest in North America

Near the end of the course, all course participants have not only mastered procedures in the laboratory, but also developed an eye for potential "chemical ecology stories" in their outdoor surroundings. Depending on geographic location, this can be the seashore, grasslands, agricultural fields, a lake, or a riverbank. Given our location, the students went into the woods surrounding our field station in teams of two for 1 or 2 h and explored. After their explorations, we gathered in the field classroom and teams reported their findings, interpretations, and research ideas. Instructor and the entire class discussed what they had seen. The instructor can elaborate on their ideas and fill in any studies that have been published on the topics the students bring up. Here is a list of topics from 1 year:

Spruce cones left by red squirrels
Gray squirrel middens
Hole in a tree: fisher den?
Old beaver sign
A scrape by mice
Chipmunk burrow
Red maple beaver chips
Mushrooms partly consumed by animals: question of toxins and avoidance
Leaves partly eaten by insects
Yellow birch odor
Browsed fern
Following a white-tailed deer, observing its food choices: red maple, wood fern,
    Canada mayflower, sugar maple
A deer fawn browsing
Mushrooms with traces of feeding by slugs and rodents
Bear rub on tree?
Deer tracks
Snowshoe hare droppings
Chipmunk droppings
Deer scats
Bat guano
Bear and deer droppings

Each of these observations or sightings has chemical ecology implications that will be discussed. The instructor will lead students to pertinent literature. Topics include, but are not limited to, feeding attractants and repellents; palatability; detoxication of plant secondary compounds; marking with secretions or excretions; chemical defense.

Amazon Forest

Needless to say, any ecosystem offers a rich variety of "chemical ecology stories" to discover. For example, on an excursion in the *Amazon forest* one can be sure to encounter many fascinating chemical relationships between organisms. Here I start with three examples concerning birds.

The hoatzin (*Opisthocomus hoazin*) is a leaf-eating bird whose strong odor earned it the name "chancho" (pig). This slow-moving bird, often in fairly open trees, appears very vulnerable to predation. Does the odor or the taste of its flesh deter predators, as H. B. Cott showed for many species of birds in a series of papers from the 1940s to the 1960s?

Sight of a yellow-headed vulture (*Cathartes* sp.) reminds us of its keen sense of smell capable of detecting a carcass, similar to the turkey vulture. The larger king vulture cues in on these scavengers with their good noses and helps them in turn open tough carcasses.

Macaws, tapirs, howler monkeys and other animals eat clay, presumably for adsorbing toxins in their food. Macaws eat more clay during the dry season when they have to rely more on seeds with their potentially harmful secondary plant metabolites. They even feed their nestlings clay before they venture outside the nest cavity (Brightsmith 2002).

We may find scent marks of the giant river otter.

What is the function of the milky latex in young branches and leaves of the *Jacaratia* sp. tree; and of the strong odor of the garlic tree (ajos chiro, or ajosquiro, *Gallizia* sp., Phytolaccaceae)?

The tangarana tree (or "torture tree," *Triplaris* sp., Polygonaceae, see Figure) is less chemically defended than other trees. Instead, it depends on ants to protect it. They live in the hollow trunk and feed on nectar produced by extrafloral nectaries, the tree's reward for being defended by the ants ("Torture tree" refers to tying someone to the tree to be bitten by the large ants as punishment).

References

Brightsmith DJ (2002) The Tambopata Macaw Project. Earthwatch Institute, Washington, DC.
Cott HB (1947) The edibility of birds. Proc Zool Soc Lond 122:371–542
Cott HB, Benson JM (1969) The palatability of birds, mainly based upon observations of a tasting panel in Zambia. Ostrich 8:335–463

Section II
Laboratory Experiments

Chapter 12
Test for Cyanogenic Compounds in Plants[1]

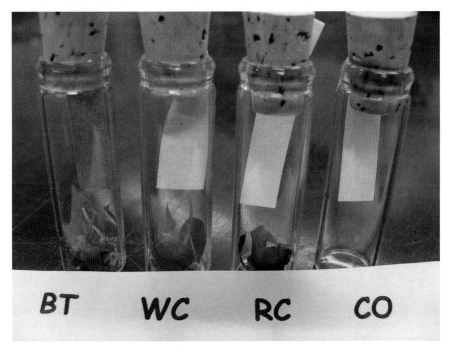

Results of a cyanide test with the Guignard sodium picrate reaction. *BT* birdsfoot trefoil, *WC* white clover, *RC* red clover, *CO* control (no plant material)

[1]Modified from an exercise kindly provided by Dr. David A. Jones, University of Florida.

Introduction

Many plants contain cyanogenic compounds. A disproportionate number of the most important cultivated human food plants are cyanogenic. Cyanogenic plants or plant parts resist microorganisms, insects, and vertebrate herbivores both in vivo and in storage and so it is possible that these plants were the successful ones during domestication because they were naturally protected against pests (Jones 1998). The production of hydrogen cyanide, HCN, termed cyanogenesis, depends on the degradation of a naturally occurring cyanogenic glycoside by an enzyme. A living plant normally keeps the substrate and the enzyme separate. They come together when the plant tissue is damaged, as by an herbivore chewing a leaf.

Plants such as birdsfoot trefoil (*Lotus corniculatus*) and white clover (*Trifolium repens*) are polymorphic with regard to cyanogenesis. This means that different plant specimens of these species can range from being completely acyanogenic to showing various degrees of cyanogenesis. This variation can be correlated with presence of herbivores such as molluscs (Jones 1962; Ellis et al. 1977; D. Jones, pers. comm. 2009). Herbivores eat the acyanogenic specimens and therefore select for cyanogenic plants in a given population. Among the samples in the figure of this chapter, birdsfoot trefoil is cyanogenic, while white clover is not. Both samples came from a rural area near Syracuse, NY.

People can suffer cyanide poisoning from poorly prepared cassava (manioc, locally also known as "yuca" = *Manihot esculenta*), a common staple in the tropics. In 2005, 27 Filipino elementary school children died and 100 persons were sickened after eating fried cassava roots (New York Times, 10 March 2005). Some organisms tolerate high levels of cyanide in their diet. The critically endangered golden bamboo lemur (*Hapalemur aureus*) of Madagascar, weighing 1.5 kg, consumes bamboo shoots that contain cyanide levels that would kill an adult person (Wright 1997). Genetic engineering can reduce the levels of cyanogenic glycosides such as linamarin and lotaustralin, in the roots of cassava, while substantial levels remain in the leaves, protecting the plants from herbivores (Pickrell 2003). The entire pathway for synthesis of the cyanogenic glucoside dhurrin has been transferred from *Sorghum bicolor* to *Arabidopsis thaliana*, conferring resistance to feeding by the flea beetle *Phyllotreta nemorum* of the Chrysomelidae family (Tattersall et al. 2001).

The test for cyanide we use is called *Guignard sodium picrate test*. It should be noted that if the sodium picrate test is positive we cannot be certain that a plant not tested previously is really cyanogenic, because the test is not sufficiently specific for cyanide. However, we know that white clover and birdsfoot trefoil are cyanogenic. [Another test for cyanide, the Feigl–Anger test is more specific for cyanide gas. It uses paper strips impregnated with copper II ethylacetoacetate ($C_{12}H_{18}CuO_6$). In the presence of cyanide gas, the color changes from pale blue-green to bright blue or purple (Feigl and Anger 1966). However, unlike the picrate test, the Feigl–Anger test is not quantitative].

Materials

1. Flat bottomed glass specimen tubes (50 × 10 mm), with corks
2. Blocks (wood or plexiglass) with 3–10 holes to hold the glass tubes
3. Strips of filter paper (Whatman # 1; 38 × 8 mm)
4. A solution of sodium picrate (Dissolve 6 g of the purest available, moist picrid acid in a liter of distilled water; stir in 50 g of anhydrous Na_2CO_3; filter and store in a dark bottle with a screw cap).
   Warning: picric acid (2,4,6-trinitrophenol) should be kept moist and in the poisons cupboard! Dry picric acid is explosive! (Many schools already have such potential bombs in their cupboards.)
5. Polyethylene bags (two per block)
6. A 100-mm Petri dish
7. A pair of forceps
8. A pair of medium or fine scissors
9. A glass rod (5 mm diameter)
10. Sulfur-free toluene in a dropping bottle with ground-glass stopper/pipette
11. Filter paper (approx. 120 mm × 120 mm, or 90-mm disks)
12. Plants to be tested: Birdsfoot trefoil (*Lotus corniculatus*) or apple seeds or almonds

Method

Basic Guignard Sodium Picrate Test

We will test birdsfoot trefoil, *Lotus corniculatus*, along with other legume species (and possibly other plants such as St. Johnswort) for comparison.

1. Prepare as many "picrate papers" as you need. Soak the papers in a puddle of sodium picrate in the Petri dish. Lift the papers so that they are arranged in a circle round the edge of the dish.
   Always use freshly prepared and moist picrate papers. Some publications suggest to prepare the picrate papers beforehand and use them dry. This is bad advice because dry papers can give false positive results.
2. Each team uses a set of seven vials: one for 3 leaves of birdsfoot trefoil (about 25 mg of material); one for 6 leaves of the same species; one each for 3 and 6 leaves of white clover (*Trifolium repens*); and one each for 3 and 6 leaves of red clover (*Trifolium pratense*). Finally, one vial is a blank, with no plant material at all.
3. Number the vials and record their contents. Use leaves nearest the growing point of a single stem. Do not include flowers or flower buds! Use the glass rod to push the leaves to the bottom of the tube. Wash and dry the rod between tests.

4. Add three drops of toluene. Use a fume hood. Toluene can be toxic if inhaled in large amounts and for prolonged time. The amount and exposure time in this experiment are both very small. Toluene is a solvent. It damages the cuticle, permitting passage of hydrogen cyanide to the outside.
5. Using a cork of the correct size as an aid, lift a strip of picrate paper out of the Petri dish, dab off the excess picrate solution on a filter paper, and place the strip in the tube holding it in place above the plant material by means of the cork. Do not let any of the paper touch the plant material, since we test for airborne HCN. Stop the paper from protruding above the end of the tube. (You will get a yellow finger, but it soon will wear off. The test also works with the paper protruding, as shown in the illustration.)
6. Repeat for the other samples.
   Note:
   (a) The position of the block
   (b) The plant material (or control) which each tube contains
7. Cover the tubes and the block with two polyethylene bags and incubate for one hour or more.

Results

- After one or more hours, score the tubes for color change of the paper from yellow to red. Decide which plants are cyanogenic. Use the blank for comparison.
- Write your one-page report in the usual format: Title, author, statement of problem, method, results, discussion, references. An illustration, if needed, will go into an appendix.

Sampling of Plant Populations for Frequencies of Cyanogenic Specimens (Polymorphism)

In this second part, we will test a number of specimens of the same species in each of several locations for variation of cyanogenesis in a population. We will use *Trifolium repens* (white clover) or *Lotus corniculatus* (birdsfoot trefoil). To collect the plants in a systematic fashion, we lay out a grid. Sticks mark the ends of rows (transects) on a lawn or in a field. We start with one plant, then pick one plant 1 m along the transect, and so forth. If no plant of interest exists at that point, the closest plant is chosen, and the next one will be 1 m from that, and so on.

Collect at least ten plants from one site, then ten each from two different sites, such as different lawns. Pick sites that differ ecologically, such as dry vs. wetter areas.

Test all plants as described in the basic experiment. Keep them separate by site of origin. Do the percentages of cyanogenic plants differ between the sites? If so, how do you explain the differences?

Quantitative Differences in Cyanogenesis Between Plants

Besides being either acyanogenic or cyanogenic, plant specimens of *Lotus corniculatus* often also differ in their degree of cyanogenesis (polymorphism). The sodium picrate test is sensitive to these quantitative differences: The resulting color can vary from light orange to deep brick-red (Jones 1962).

Obtain a color chart with variations of yellow, orange, and red. Match the color each picrate paper after reaction with plant gases against the chart. Classify the levels of cyanogenesis from 0 to 6. Zero means no reaction, i.e., the plant is acyanogenic. Six is the highest level, indicated by a deep brick-red.

References

Brinker AM, Seigler DS (1989) Methods for the detection and quantitative determination of cyanide in plant materials. Phytochem Bull 21:24–31

Conn EE (1979) Cyanide and cyanogenic glycosides. In: Rosenthal GA, Janzen DH (eds) Herbivores: Their interaction with secondary plant metabolites. Academic, New York, NY

Cooper-Driver GA, Swain T (1976) Cyanogenic polymorphism in bracken (*Pteridium*) in relation to herbivore predation (*Schistocerca gregaria*). Nature 260:604

Ellis WM, Keymer RJ, Jones DA (1977). On the polymorphism of cyanogenesis in *Lotus corniculatus* L. VIII Ecological studies in Anglesey. Heredity 39:45–65

Feigl F, Anger V (1966) Replacement of benzidine by copper ethylacetoacetate and tetra base as spot-test reagent for hydrogen cyanide and cyanogen. Analyst 91:282–284

Jones DA (1962) Selective eating of the acyanogenic form of the plant *Lotus corniculatus* L. by various animals. Nature 193:1109–1110

Jones DA (1998) Why are so many food plants cyanogenic? Phytochemistry 47:155–162

Jones DA, Parsons J, Rothschild M (1962) Release of hydrocyanic acid from crushed tissues of all stages in the life-cycle of species of the Zygaeninae (Lepidoptera). Nature 193:52–53

Pickrell J (2003) Cyanide on the side. http://sciencenow.sciencemag.org.5/30/2003

Tattersall DB, Bak S, Jones PR, Olsen CE, Nielsen JK, Hansen ML, Hoej PB, Moeller BL (2001) Resistance to an herbivore through engineered cyanogenic glucoside synthesis. Science 293:1826–1828

Wright PC (1997) Behavioral and ecological comparisons of neotropical and Malagasy primates. In: Kinzey WG (ed) New world primates: Ecology, evolution and behavior. Aldine Transaction, Piscataway, NJ, p 127ff

Chapter 13
Herbivory and a Simple Field Test for Total Phenolics in Trees

Some northern trees and phenolics reactions with ferric chloride. The phenolics stains are shown on cross sections of twigs in the *middle two rows*

D. Müller-Schwarze, *Hands-On Chemical Ecology: Simple Field and Laboratory Exercises*,
DOI 10.1007/ 978-1-4419-0378-5_13, © Springer Science+Business Media, LLC 2009

Many deciduous and coniferous trees contain varying levels of phenolics, including tannins. "Blackwater rivers" in forested areas of the world, such as the Amazon basin or the southern United States, carry phenolics (tannins) leached out of trees. Such rivers differ in plant and animal life from "whitewater rivers."

The biological functions of such plant secondary metabolites (PSMs) have been debated for a long time. They often have antimicrobial functions, but also serve as repellents and feeding inhibitors against herbivorous insects and vertebrates, notably birds and mammals. Animals have evolved many mechanisms to cope with phenolics in their diet. These start with food processing. For instance, beavers consume experimental sticks of the phenolics-rich witch hazel only after leaving them in the water for 2–3 days, apparently to leach out unpalatable compounds (Müller-Schwarze et al. 2001). Many birds and mammals eat clay to adsorb phenolics so they never will be absorbed in the intestines. If they are taken up in the blood stream, such PSMs will eventually be rendered harmless by oxidation and other processes, followed by conjugation, in the liver. They then will be excreted in the urine.

The levels of phenolics vary with plant parts and season. Most valuable parts such as buds, flowers, and catkins are more heavily defended. For the winter, more phenolics are translocated to the bark. This is thought to intensify defense at a time when the tree is dormant and cannot respond to herbivore damage by wound healing and regrowth. Among the phenolic glycosides, salicin (saligenin glycoside) and salicortin occur in the bark of willow (*Salix* spp.) and "poplar" (*Populus* spp.), tremulacin in the bark of *P. tremula* and *P. tremuloides*. The heartwood of *P. tremuloides* contains tremulone and related compounds.

In the field, a quick test for the general level of total phenolics in a plant can be the starting point for a more detailed quantitative analysis of the nature and amounts of specific compounds. Various oxidation-reduction methods have been exploited to analyze total phenolics in plant extracts. Many, but not all, phenols form complexes with ferric chloride ($FeCl_3$). Ferric chloride gives a color reaction with phenolic compounds. While the phenolate ion is oxidized, the ferric ions are reduced to the ferrous state.

To gain an impression of the relative concentrations of total phenolics in a number of common northeastern deciduous and coniferous trees, we will apply a ferric chloride solution to freshly cut twigs of trees and observe color reactions in vivo. The species chosen are known to be browsed to varying degrees by wildlife such as deer, porcupines, cottontails, beavers, voles, or ruffed grouse.

Colors will indicate the levels of phenolics, ranging from light green to blackish purple. Also, different compounds give different colors. For example, in aqueous solution, ferric chloride yields purple for phenol, blue for *p*-cresol, green for pyrocatechol, and red for pyrogallol. (If ferric chloride is dissolved in methanol instead of water, all four compounds stain green; Snell and Ettre 1973.) Here we will have to deal with tree species whose multiple phenolic compounds will mask each other's color.

Materials Needed

1. Boughs of local deciduous and coniferous trees. If possible, include juvenile and adult growth forms of the same species.
2. Ferric chloride ($FeCl_3$) solution
3. Eyedropper
4. Clippers

Method

Herbivory Score

First, before you perform the phenolics test, rank the trees to be tested according to their known use by wildlife such as deer, beaver, cottontails, voles, porcupines, or others. Draw on your readings, other courses taken, and personal experience. As antiherbivore compounds occur in all parts of the tree, albeit at different concentrations, consider all damage, whether it occurs to leaves, twigs, bark, buds, etc. Phenolics levels vary with the seasons. For instance, in autumn these compounds are being translocated to plant parts that need protection in winter. So, even if your course takes place during the growing season, consider also winter damage. Give each tree species a score from 0 to 3 (Data Sheet 13.1).

Assign a rating to each tree species below:

0 Not known to be used by wildlife
1 Browsed very little or only by one or few species
2 Regularly, but moderately browsed
3 Heavily used, at least at times
? Do not know

Data Sheet 13.1 Levels of browsing for various trees

Species	Scientific name	Level of browsing
Deciduous trees		
Apple	*Malus sylestris*	
American basswood	*Tilia americana*	
American beech	*Fagus grandifolia*	
Hawthorn	*Crataegus* sp.	
Eastern hop hornbeam	*Ostrya virginiana*	
Red maple	*Acer rubrum*	
Sugar maple	*Acer saccharum*	
Witch-hazel	*Hamamelis virginiana*	
Northern red oak	*Quercus rubra*	
Quaking aspen	*Populus tremuloides*	

(continued)

Data Sheet 13.1   (continued)

Species	Scientific name	Level of browsing
European buckthorn	*Rhamnus cathartica*	
Willow	*Salix* sp.	
Conifers		
Larch	*Larix* sp.	
Norway spruce	*Picea abies*	
Eastern white pine	*Pinus strobus*	
Scots pine	*Pinus sylvestris*	

Test for Phenolics

You will be given a ferric chloride solution of 0.1 M $FeCl_3$ in 0.1 M HCl (5 g anhydrous ferric chloride in 100-ml water). It appears yellow and clear. First, cut an 8–10-cm-long section of a 1-cm thick twig with clippers to obtain a fresh cross-sectional surface.

Using the eyedropper, spread one drop of the ferric chloride solution over the entire surface of one end of the sample stick. A stain will develop immediately. Use the other end as untreated control. Cut the stick in half and compare stained end and control end side by side. Record the color, intensity, and distribution of the stain that appears within the first 30 s:

Rate the intensity of the stain, from 1 (pale hue) to 3 (blackish purple).

Sketch the distribution of the coloration across the cross section of the twig: concentric rings, only at cambium, only in center (heartwood), all over, etc. (Data Sheet 13.2).

Herbivory: Phenolics Correlation

- Using a spreadsheet program, correlate your ratings of the phenolics levels with the level of herbivory.
- Correlate levels of herbivory and phenolics graphically, with phenolics score on *x*-axis, and herbivory score on *y*-axis.
- Write down your conclusion. To what degree are phenolics levels correlated with levels of herbivory? Are there outliers that do not fit a general trend?
- Also, examine the leaves you have been working with. How much damage by herbivores – primarily insects, but also mammals – for each species? Keep in mind that the few boughs we have brought to the laboratory are only an infinitesimal sample of the plant universe out there. Nevertheless, does your observation jibe with the "estimated guesses" of relative levels of herbivory for each plant species?

Data Sheet 13.2 Ferric chloride test for phenolics: Results for some deciduous and coniferous trees

Species	Color heartwood/ cambium	Intensity	Pattern of stain (sketch)	Overall rating (0–3)	Herbivory score (0–3)	Remarks
Deciduous trees						
Apple						
Quaking aspen						
Willow						
Beech						
American basswood						
European buckthorn						
Hawthorn						
Eastern hop hornbeam						
Sugar maple						
Red maple						
Northern red oak						
Willow						
Witch hazel						
Coniferous trees						
Hemlock						
Larch						
Douglas fir						
Norway spruce						
White pine						
Scots pine						

Growth Stage Differences

Chapter 15 will address the different *growth forms* and *induced defenses* in detail. If time is available, growth forms can be studied in the context of the current exercise as well.

Gross morphology differences. Examine the twigs provided by the instructor. What is the branching pattern of each type? Do leaf sizes differ?

Chemical Reactions. Perform the same phenolics test with twigs of both juvenile and adult growth forms of the same species. Use whatever species are available in your area: aspen, cottonwood, red oak, willow, basswood, ash, or others. Compare the color reaction in juvenile shoots with those in shoots of the adult growth form.

References

Müller-Schwarze D, Brashear H, Kinnel R, Hintz KA, Lioubomirov A, Skibo C (2001) Food processing by animals; do beavers leach tree bark to improve palatability? J Chem Ecol 27:1011–1028

Snell FD, Ettre LS (eds) (1973) Encyclopedia of industrial chemical analysis, vol 17. Interscience, New York, NY, pp 16/17

Waterman PG, Mole S (1994) Analysis of phenolic plant metabolites. Blackwell Scientific, Oxford

Chapter 14
Radial Diffusion Assay for Tannins

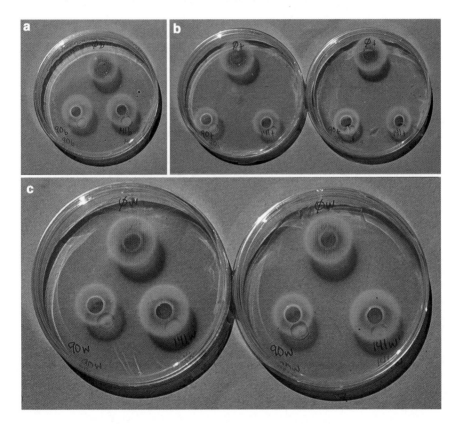

Results of a radial diffusion test for tannins in buried acorns of red oak. The three wells in each Petri dish contain extract from acorns from the same collection, but buried for different lengths of time: 0 days (kept in cold storage) at *top* in Petri dish; 91 days in the ground at *lower left* in Petri dish; 141 days in ground at lower right of Petri dish. (**a**) Extracts of base of acorns; (**b**) extracts of tips (containing embryo) of acorns; (**c**) extracts of whole acorns. Diameter of rings parallels tannin levels

D. Müller-Schwarze, *Hands-On Chemical Ecology: Simple Field and Laboratory Exercises,*   81
DOI 10.1007/ 978-1-4419-0378-5_14, © Springer Science+Business Media, LLC 2009

Tannins are polyphenols familiar to most of us as astringent compounds in red wine (originating in skins, seeds, and stems of grapes), tea leaves, and unripe fruit. In trees, they occur in bark, buds, and fruits, and are found in cell vacuoles or surface waxes. Tannins are assumed to defend plants against microorganisms, insects, and vertebrate herbivores. These brownish or yellowish compounds are used for tanning and dyeing. Tannins precipitate with proteins, the basis for tanning leather.

In this assay, the interaction of tannins with protein in an agar gel is quantified. The insoluble precipitates form rings around an origin. The diameter of the rings is proportional to the tannin amounts present. It needs to be added that not all tannins bind to proteins, and not all precipitates are insoluble. The method of this exercise follows the Tannin Assay as described by Hagerman (1987).

The material analyzed in this exercise can be known amounts of tannins such as "tannic acid" (a mixture of tannins), or plant extracts with unknown amounts of tannin. This exercise is written for tests with plant materials whose tannin level we wish to determine. These can be bud scales of trees such as aspen, or different parts of acorns (tip vs. base).

Materials Needed

1. Petri dishes (about 30 or more)
2. Agarose (6 g or more, depending on scale of experiment)
3. Bovine serum albumin (600 mg or more)
4. 4-mm cork borer
5. Plant extract
6. Pipettes
7. Scale
8. Spatula
9. Measuring tape or small ruler
10. Refrigerator
11. Incubator
12. Hot plate stirrer

Procedure

Prepare Agar–BSA–Plate as Medium

For five plates, dissolve 1 g of agarose in 100 ml water, resulting in a 1% (w/v) solution. To buffer, add 50 mM acetic acid, 60 µM ascorbic acid, adjusted to pH 5.0 with sodium hydroxide. Bring slowly to boil (on hot plate stirrer) to completely dissolve the agarose. Cool to 45°C. (Warning: Failure to let the mixture completely cool will denature the protein). Add 100 mg bovine serum albumin (BSA) [to 0.1%

(w/v)] as protein source. Let it dissolve on its own for several minutes (Stirring will result in clumping). Once dissolved, use stir bar on "low." Do not add heat! Label Petri dishes ahead of time. Pipette 20 ml of the agarose-BSA mix into each 8.5-cm diameter Petri dishes. Let them cool and solidify before moving them. Leave out overnight to evaporate some moisture. Store in refrigerator at 4°C until use.

Material to be Analyzed

Known amounts of commercial tannic acid or plant materials. For this exercise, we use the latter.

Prepare plant extract. Grind material (acorn parts, bud scales, bark, or leaves) into fine powder. Extract 100 mg material (ground acorn tips or bases, or powdered bud scales) in 0.5-ml MeOH/H_2O (50:50) [or a 70% acetone–MeOH mixture] for 1–3 h at room temperature. Pipette off supernatant.

Incubate Plates

Cut wells of 4-mm diameter with a cork borer. Make three or four wells per dish. Holes should be at least 1.5 cm apart. Place a sample of 8 μl in each well with a micropipette. Seal dishes with parafilm. Incubate at 30°C for 96–120 h. The higher the tannin concentrations, the longer it takes to reach equilibrium. The size of the diffusion rings will stabilize after about 5 nights.

In addition to the samples to be tested, run several known concentrations of a tannin, or tannin mixture such as tannic acid.

Results

Measure Tannins

1. Measure two diameters (at right angles to each other) of the rings around the wells and average them. Square the diameter. The square of the diameter is proportional to the amount of tannin present (Hagerman 1987). The diffusion area A for each ring is computed by $A = \pi r^2$. Subtract the area of the well for accuracy.
2. Test for significance of differences of ring sizes for the five replicates of each treatment by ANOVA and/or pairwise comparisons by *t* test:
   - Untreated acorns: tip vs. base
   - Acorns in soil for 90 days: tip vs. base
   - Acorn tip: untreated vs. 90 days in soil
   - Acorn base: untreated vs. 90 days in soil

Previous Results

The author and his students analyzed tannin levels in tips and bases of acorns that had been buried in the ground for 0, 90, and 141 days, starting from 28 October. Tannin in the tips decreased from 0 to 90 days, and then increased again until 141 days. The same was observed for extracts of whole acorns. The levels in the base of the acorn did not change at all (see figure). (The buried acorns were also ca. 30% heavier than the ones kept in cold storage). How do you interpret these results?

References

Hagerman AE (1987) Radial diffusion method for determining tannin in plant extracts. J Chem Ecol 13:437–449

Jakubas WL, Gullion, GW, Clausen TP (1989) Ruffed grouse feeding behavior and its relationship to secondary metabolites of quaking aspen flower buds. J Chem Ecol 15:1899–1918

Peng S, Jay-Allemand C (1991) Use of antioxidants in extraction of tannins from walnut plants. J Chem Ecol 17:887–896

Schmidt KA, Brown JS, Morgan RA (1998) Plant defenses as complementary resources: a test with squirrels. OIKOS 81:130–142

Smallwood PD, Peters WD (1986) Grey squirrel food preferences: the effects of tannin and fat concentration. Ecology 67:168–174

Chapter 15
Chemically Induced Defenses in Phytoplankton

Kimberly L. Schulz

Introduction

Many organisms, including some phytoplankton and zooplankton, are phenotypically plastic, and respond to predators by changing their shapes – a phenomenon known as "inducible defense." For example, some rotifers grow large spines in the presence of their predators, and some clones of the common crustacean *Daphnia* develop protective "neckteeth" protrusions when their predator, the insect larva *Chaoborus* (the phantom midge), is abundant. Many predators and grazers are limited by the size of food they can consume, and individuals with these spines and bumps have reduced predation risk.

One of the most interesting aspects of these shape changes is that they are a morphometric response to the presence of compounds released either by the predator itself or by the predator actively feeding. For these inducible defenses, the prey does not exhibit the spines when the predator is absent, but develops the defensive morphology in response to a water-borne cue from the predator. Repeated studies have demonstrated that many changes in shape of zooplankton and phytoplankton do not require the presence of the predator, but can be induced just by adding water from an aquarium or lake where the predator is present. So, the changes we observe in lakes, where zooplankton and phytoplankton often become spinier when a predator becomes abundant, are not necessarily due to the predators eating all the nonspiny individuals and leaving only the spiny members of the population, but also can be caused by inducible shape changes.

It is generally believed that there are costs associated with maintaining spines and other inducible defenses, and in fact some known costs for zooplankton that are spiny include reduced growth and reproduction rates, and increased energetic costs of swimming. Many of the compounds responsible for the signal that induces shape changes in plankton have not yet been isolated.

One example of an inducible defense that is quick (several days) and easy to observe in the lab is an increase in colony size and spine length in the common green phytoplankton, *Scenedesmus* (Chlorococcales, Chlorophyta), when exposed to water in which the common zooplankton grazer, *Daphnia* (Cladocera, Crustacea), is cultured. Researchers have demonstrated reduced grazing rates of *Daphnia* on

D. Müller-Schwarze, *Hands-On Chemical Ecology: Simple Field and Laboratory Exercises,*
DOI 10.1007/ 978-1-4419-0378-5_15, © Springer Science+Business Media, LLC 2009

larger colonies of *Scenedesmus* compared with small colonies or single cells. Costs suggested for colony formation in *Scenedesmus* include reduction in photosynthetic rates and increased likelihood of sinking out of the surface waters of the lake. Although the chemical structure of the compounds responsible for colony stimulation in *Scenedesmus* is not known, research has shown that the responsible molecules are organic, moderately lipophilic, nonvolatile, and of small mass (<500 Da). These infochemicals are not derived from the algae themselves, because algal homogenates do not induce colony formation, and they are also not released by starving *Daphnia* or those eating polystyrene beads. The chemicals that induce the colony formation are released by *Daphnia* and several other common zooplankton (the rotifer *Brachionus calyciflorus*, the copepod *Eudiaptomus gracilis*, and the small cladoceran *Bosmina longirostris*), as they actively feed on phytoplankton.

In this exercise, we test the effects of infochemicals present in cultures of *Daphnia* grazing on the green alga *Scenedesmus* on inducible colony formation in the alga. We asked the question:

1. Does the green alga *Scenedesmus* increase colony size in response to the presence of filtered water from a *Daphnia* culture?

Some additional questions that could be asked with this experimental setup are as follows:

1. Does the chemical cue causing induction persist over time?
2. Is the response of colony formation concentration dependent?
3. Does water from cultures of other zooplankton induce colony formation?

Materials Needed

1. Culture of *Scenedesmus* phytoplankton: *Scenedesmus* can be obtained from culture collections or scientific supply companies. Many species in the genus *Scenedesmus*, including *S. acutus*, *S. subspicatus*, and many but not all clones of *S. obliquus* will exhibit colony induction.
2. Culture of *Daphnia*: *Daphnia* is a common zooplankton species (the water flea) that can be found in many lakes and ponds. It can also be purchased from scientific supply companies. *Daphnia* is easy to grow in lake water with a supply of algae in any type of jar (>500 ml).
3. Beakers or flasks (>200 ml) for growing algae.
4. Labeling tape.
5. Permanent marker.
6. Parafilm or aluminum foil to cover the algal cultures.
7. Filtered lake water or algal growth medium for growing algae.
8. Laboratory space to grow phytoplankton and zooplankton for a week. This could be an incubator (with lights, set at a temperature between 15 and 25°C), or a room that does not get too hot either equipped with some grow lights or aquarium lights for the phytoplankton or near a bright window (or in a greenhouse).

9. Pipettes for dispensing known quantities of algae and culture water.
10. Filtration apparatus.
11. Filters (<10 μm; preferably <2 μm polycarbonate filters or glass fiber filters).
12. Containers for collecting filtered *Daphnia* culture and control water.
13. Compound microscope.
14. Microscope slides or counting chambers.

Procedure

Preparation of Daphnia Culture

Obtain *Daphnia* from a local lake or pond or from a scientific supply house. Grow them in a clean glass jar filled with either filtered lake water or with a culture medium recommended by the supply company.

Preparation of Scenedesmus Culture

Obtain *Scenedesmus* from a scientific supply company or algal culture collection. Raise them on a standard culture medium recommended by the supply company.

Initial Sampling of Scenedesmus and Setup of Experimental and Control Beakers

First collect a sample of the initial *Scenedesmus* to determine the colony size before treatments are applied.

Stir the initial sample gently to mix it and put a drop onto a microscope slide. Under a compound microscope count the number of cells in the first 50 colonies encountered. If the counting is all to be done in a single class, then this sample could be preserved (Lugol's solution or glutaraldehyde) and counted later. If the microscope is sufficiently powerful, drawings of several colonies can be made to estimate roughly the spine sizes on the cells.

The *Scenedesmus* culture should then be divided between control and experimental treatment beakers or flasks. If necessary, the *Scenedesmus* culture can be diluted with culture media to ensure sufficient material for the experiment.

We have every student set up one replicate of each treatment. The beakers are labeled with the treatment. Culture medium is used as "control" water and filtered water from the *Daphnia* culture (to remove *Daphnia* and *Scenedesmus*) is used as the "experimental" water. Five milliliters of the appropriate water is added to each beaker.

Each day for at least 3 days (optimally for a week), students add 5 ml more of filtered control or experimental water to the *Daphnia* cultures.

Final Sampling

After 3 days to 1 week (longer would be fine, but would necessitate more daily application of experimental and control water) stir the samples gently to mix. For each replicate of each treatment place a drop under the compound microscope and record the number of cells in the first 50 colonies encountered. Again, observe the spines if the microscope's resolution permits (Data Sheet 15.1).

Data Sheet 15.1 Number of cells per colony for (1) initial samples of *Scenedesmus*, (2) experimental *Scenedesmus* (filtered water from a *Daphnia* culture consuming *Scenedesmus*), and (3) control *Scenedesmus* (filtered culture water added). An additional column is provided if an extra treatment was performed. On each line record the number of cells per colony for the sample indicated by the column heading

| Colony # | Initial (week 1) | Final (week 2) | | |
	Initial	*Daphnia* water	Control	Extra treatment
1				
2				
3				
4				
5				
6				
7				
8				
9				
10				
11				
12				
13				
14				
15				
16				
17				
18				
19				
20				
21				
22				
23				

(continued)

Data Sheet 15.1 (continued)

Colony #	Initial (week 1) Initial	Final (week 2) Daphnia water	Control	Extra treatment
24				
25				
26				
27				
28				
29				
30				
31				
32				
33				
34				
35				
36				
37				
38				
39				
40				
41				
42				
43				
44				
45				
46				
47				
48				
49				
50				
Average/SD				

Results

Tabulate the number of cells in the *Scenedesmus* colonies for the initial sample, the control (no *Daphnia* culture water) algal culture, and your experimental culture(s). All of the data from one lab section can also be combined to give replicates. Calculate the mean and standard deviation of colony size for the initial, control, and all treatments. Plot histograms of the size distribution of colonies (1–16 cells) of the different treatments and do simple statistical tests to compare the mean sizes (e.g., paired *t* tests) or the distributions.

Write Report

Write up results and discuss finding. You may want to suggest other types of experiments to try to isolate the chemical responsible for inducing the morphological changes in *Scenedesmus*.

Modifications for Additional Treatments

Three potential easy additions to this experiment were suggested (and many more are possible – students may be able to suggest their own extra treatments). The additional steps necessary for these are:

1. Does the chemical cue causing induction persist over time?
   Here the only necessary modification is to let the "experimental" water (filtered *Daphnia* water) sit for a known longer period of time before addition to the *Scenedesmus*.
2. Is the response of colony formation concentration dependent?
   Here the "experimental" water can be diluted serially with culture medium to produce different concentrations to be added to the *Scenedesmus*.
3. Does water from cultures of other zooplankton induce colony formation?
   Here water from other types of zooplankton cultures would be filtered and added to the *Scenedesmus*.

References

van Donk E (2007) Chemical information transfer in freshwater plankton. Ecol Informat 2:112–120

Hessen DO, van Donk E (1993) Morphological changes in *Scenedesmus* induced by substances released from *Daphnia*. Arch Hydrobiol 127:129–140

Lampert W, Rothhaupt KO, von Elert E (1994) Chemical induction of colony formation in a green alga (*Scenedesmus acutus*) by grazers (*Daphnia*). Limnol Oceanogr 39:1543–1550

Lürling M, van Donk E (1997) Morphological changes in *Scenedesmus* induced by infochemicals released *in situ* from zooplankton grazers. Limnol Oceanogr 42:783–788

Lürling M (1999) Grazer-induced coenobial formation in clonal cultures of *Scenedesmus obliquus* (Chlorococcales, Chlorophyceae). J Phycol 35:19–23

Verschoor AM, van der Stap I, Helmsing NR, Lürling M, van Donk E (2004) Inducible colony formation within the Scenedesmaceae: Adaptive responses to infochemicals from two different herbivore taxa. J Phycol 40:808–814

Chapter 16
Induced Defense: Herbivory on Juvenile vs. Adult Growth Stages of Trees

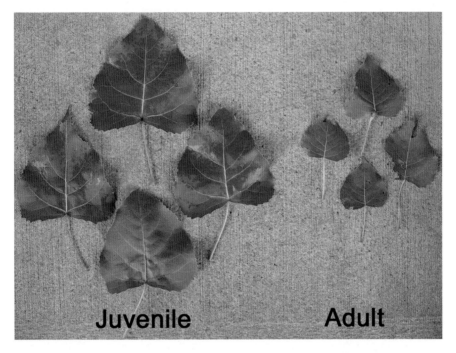

Two different growth forms of leaves of Eastern cottonwood (*Populus deltoides*). On *left*: juvenile growth type (regrowth after cutting down of young tree). On *right*: adult growth form. Note that the adult form has more herbivore damage

D. Müller-Schwarze, *Hands-On Chemical Ecology: Simple Field and Laboratory Exercises*,   91
DOI 10.1007/ 978-1-4419-0378-5_16, © Springer Science+Business Media, LLC 2009

Plants contain "plant secondary metabolites" (PSMs) that confer resistance against herbivores and pathogens. The largest classes of PSMs are phenolics, terpenoids, and alkaloids. We distinguish *constitutive* and *induced* chemical defenses. The first occur in an entire plant or some of its parts as normal constituents, the latter are formed in response to herbivory or some other trauma to the plant. In some cases mere clipping induces defenses, in others additional factors, such as those in saliva of a herbivore are essential.

Some northern trees such as green alder or paper birch respond to browsing by mammals such as snowshoe hares by growing adventitious shoots that are richer in PSMs than older twigs (Bryant 1981). These new shoots represent the *juvenile-type growth* form, while the older twigs on the tree or shrub are of the *adult-type growth* form. The fresh shoots with induced PSMs are avoided by hares for 1–3 years. In our area, quaking aspen, *Populus tremuloides*, that is growing back after cutting by beavers is more heavily defended during its first few years of regrowth (Basey et al. 1990).

Materials Needed

1. Boughs of deciduous trees typically produce adventitious shoots when browsed or cut down. These include species of the genera *Populus*, *Salix*, *Tilia*, and others. In our experience, Eastern cottonwood (*P. deltoides*) and American basswood (*T. americana*) give good results in terms of species similarities and differences.
2. A small ruler for each student (10–20 cm).

Procedure

Leaf Sizes

To demonstrate differences between juvenile and adult growth forms, we use Eastern cottonwood, *Populus deltoides*, as example. (The exact species will differ by region and locality, of course). You will be provided with two (or three) batches of leaves collected locally. One is from a mature tree and displays the regular *adult-type growth form*. A second sample represents the *juvenile-type growth form*. It had regrown from rootstock on a stump after browsing by animals or being cut down by humans. Often we find such new shoots in a ditch after mowing or cutting by road maintenance crews. (There may be available a third type of leaves: regrown on a mature tree after branches had been clipped.)

1. *Gross morphology differences.* Examine the twigs provided by the instructor. What is the branching pattern of each type?

2. *Leaf size*. Measure the sizes of ten leaves. The leaves of Eastern cottonwood are easy to measure, as they are nearly triangular. Measure baseline and height with a small ruler. Use formula for the area of a triangle to arrive at the area of the leaf.

Results (Leaf Size Differences)

- Tabulate and compare in Data Sheet 16.1. Use a spreadsheet to compute leaf areas.

Insect Herbivory

1. Count insect holes: *Insect herbivory*. Each student examines ten juvenile leaves and ten adult form leaves, and counts insect holes in all of them. Since you work in teams of 2 or 3, each student handles only a few leaves of each type. (If time is limited, a smaller number might also suffice, as you will pool the data for the entire class. Obviously, the larger the class, the fewer leaves each student will have to study). Count holes in the leaf and damage at the edge separately, as insect species differ in their feeding patterns, and affect tree species differently.
2. Construct two tables as given in Data Sheets 16.2 and 16.3.
3. Relate numbers of insect holes to leaf area. Using a spreadsheet, compute numbers of insect holes per area. Since juvenile leaves are larger and have fewer insect holes, this will show any differences in insect herbivory even better.

	Insect holes per 50 cm^2 leaf area	
Leaf #	Juvenile growth form	Adult growth form
1		
2		
3		
4		
5		
6		
7		
8		
9		
10		
Total		
Mean		

Data Sheet 16.1   Areas (cm²) of juvenile-type and adult-type leaves of Eastern cottonwood

Leaf #	Width (cm)	Length (cm)	Area (cm²)
1			
2			
3			
4			
5			
6			
7			
8			
9			
10			
Total			
Mean			

Data Sheet 16.2   Numbers of insect holes in juvenile- and adult-form leaves

Cottonwood juv	Cottonwood adult	Basswood juvenile	Basswood adult

Data Sheet 16.3   Numbers of notches at leaf edges

Cottonwood juv	Cottonwood adult	Basswood juvenile	Basswood adult

Results (Insect Herbivory)

(a) Statistics: Compute statistical significance of difference in insect damage per leaf area. Use *t* test. You may find enormous differences.
(b) Graph data.
(c) As an optional shortcut that saves time, you can lump the insect damage into three categories: intact leaves, leaves with minor damage, and leaves with heavy damage. The last two categories can be combined into "leaves with damage." Tabulate and compare by χ^2 test. Use Data Sheet 16.4. You end up with a table as given in Data Sheet 16.4.

Smell Test

Try to smell a difference between the two types of leaves: Place each of five juvenile leaves in five separate small paper bags with a sniffing hole. Put three adult leaves in each of five identical paper bags. (The larger number is to compensate for the smaller leaf size.) Each student sniffs each bag with the invisible leaf. Since we know that juvenile leaves are supposed to be better defended, for each bag we have to answer the question: Is this leaf of the juvenile or adult type? (Data Sheet 16.5)

Data Sheet 16.4 Overall insect herbivory

Growth form	# Intact leaves	# With minor damage	# With major damage	Total # of all damaged leaves
Juvenile				
Adult				

Data Sheet 16.5 One student's choices in sniff test

Leaf #	Adult (a) or juvenile (j)?	Correct type (to be filled in after choice has been made)	Score: + or − (correct or incorrect)
1			
2			
3			
4			
5			
6			
7			
8			
9			
10			
Total number of correct choices			

Data Sheet 16.6 Compilation for results of entire class

Student	Correct (+) or incorrect (−) choices made, by individual leaf number										
	Leaf Nr.1	2	3	4	5	6	7	8	9	10	Total correct choices
A											
B											
C											
D											
E											
F											
G, etc.											
Total number of correct choices out of ..... choices made: % Correct:											

Results (Smell Test)

- Test significance of result by means of a one-sample t test against an expected random average of five correct choices.
- Discuss possible reasons for differences and any caveats.

For the whole class, compile data in a table as given in Data Sheet 16.6 (some data entered as example).

Optional: *Chemical Reactions*. Perform the *phenolics test* (see Chap. 13) with twigs of both juvenile and adult growth forms of the same species. Use whatever species are available in your area: aspen, cottonwood, red oak, willow, basswood, ash, or others. Compare the color reaction in juvenile shoots with those in shoots of the adult growth form.

References

Basey JM, Jenkins SH, Busher PE (1990) Food selection by beavers in relation to inducible defenses of populus tremuloides. Oikos 59:57–62

Bryant JP (1981) Phytochemical deterrence of snowshoe hare browsing by adventitious shoots of four Alaskan trees. Science 213:889–890

Fredrickson EL, Estell RE, Remmenga MD (2007) Volatile compounds on the leaf surface of intact and regrowth tarbush (*Flourensia cernua* DC) canopies. J Chem Ecol 33:1867–1875

Kessler A, Baldwin IT (2002) Plant responses to insect herbivory: The emerging molecular analysis. Ann Rev Plant Biol 53:299–328

Chapter 17
Jasmonic Acid Effect on Plant Volatiles (or How to Make a Fern Smell Like a Rose)

Insects feeding on leaves also trigger chemical responses by the plant that result in either less herbivory or attracting parasites or predators of herbivorous insects. *Volicitin* [*N*-(17-hydroxylinolenoyl)-L-glutamine] in the saliva of caterpillars of the beet armyworm (*Spodoptera exigua*) triggers a cascade of C_{18} compounds. This cascade leads to the production of jasmonic acid (or jasmonate). Jasmonate (or methyl jasmonate) in turn activates genes that function in production and release of plant volatiles, particularly terpenes. These volatiles deter herbivores, and also attract insects that parasitize or prey upon the herbivorous insects that feed on the plant. In this tripartite regulatory mechanism, the plant "calls for help" when attacked. The relationships can be complex: The grass Tall fescue (*Lolium arundinaceum*) upregulates defense compounds when stimulated by methyl jasmonate, but only if it does not harbor a mutualistic fungus (Simons et al. 2008).

This exercise is very simple and requires very little time. Of course, it can be expanded in several ways. We simply treat a term with jasmonate and test for terpene production by its odor.

Materials Needed

1. Fern (several fronds)
2. Jasmonic acid (methyl jasmonate)
3. About 18 vessels (beakers) per class, for keeping plants in water
4. A spray bottle

Procedure

In this simple experiment, the 15 plants (or plant parts such as fronds) are for the entire class, so that each student or team of students works with one or two specimens only. Apply a 1 mM aqueous solution (105 mg per 500 ml water) of jasmonic acid

D. Müller-Schwarze, *Hands-On Chemical Ecology: Simple Field and Laboratory Exercises,*   97
DOI 10.1007/ 978-1-4419-0378-5_17, © Springer Science+Business Media, LLC 2009

Data Sheet 17.1 Compilation of mean odor ratings of three treatments by j students

Student	Systemic JA treatment	JA sprayed on leaf	Control: No JA
1	(mean of five ratings)		
2			
3			
4			
5			
6			
7			
8			
9			
j			

(JA) to five sprigs of a plant that normally have no terpene smells, such as a fern frond. The five sprigs should be in five separate vessels ("vases"). There are two ways of treating a plant with JA: systemic (in the water it takes up) and spraying it (dissolved in ethyl alcohol) on selected leaves.

Equal numbers of plants will receive each of the three treatments: systemic, leaf only, and untreated control (no JA).

Leave the treated plants for several hours or overnight. Test for presence of terpenes by smelling the plants and comparing treated and untreated specimens.

Results

- Each observer categorizes each code-labeled plant (or treated leaf) as: (1) no smell, (2) weak smell, and (3) strong smell.
- Calculate the mean scores for each student's five ratings of the three categories. Compile these means in Data sheet 17.1.
- Test for significance of ratings of treated vs. control plants with a block-design analysis of variance. The students are the blocks (replications).

References

Cipollini DF, Sipe M (2001) Jasmonic acid treatment and mammalian herbivory differentially affect chemical defenses and growth of *Brassica kaber*. Chemoecology 11:137–143

Dicke M, Gols R, Ludeking D, Posthunus MA (1999) Jasmonic acid and herbivory differentially induce carnivore-attracting plant volatiles in Lima bean plants. J Chem Ecol 25:1907–1922

Liechti R, Farmer EE (2002) The jasmonate pathway. Science 296:649–650

Redman A, Cipollini DF, Schultz JC (2001) Fitness costs of jasmonic acid-induced defense in tomato, *Lycopersicon esculentum*. Oecologia 126:380–385

Simons L, Bultman TL, Sullivan TJ (2008) Effects of methyl jasmonate and an endophytic fungus on plant resistance to insect herbivores. J Chem Ecol 34:1511–1517

Thaler JS (1999) Jasmonic acid mediated interactions between plants, herbivores, parasitoids, and pathogens: A review of field experiments in tomato. In: Agrawal AA, Tuzun S, Bent E (eds) Inducible plant defenses against pathogens and herbivores: Biochemistry, ecology, and agriculture. American Phytopathological Society Press, St. Paul, MN

Chapter 18
Effect of Tannins on Insect Feeding Behavior

Caterpillars of the tobacco hornworm, *Manduca sexta*, in individual containers, feeding on tannin-treated laboratory chow

D. Müller-Schwarze, *Hands-On Chemical Ecology: Simple Field and Laboratory Exercises,*   101
DOI 10.1007/ 978-1-4419-0378-5_18, © Springer Science+Business Media, LLC 2009

The feeding behavior of herbivorous insects is guided by plant chemistry. There are specialists and generalists with regard to the range of plant species they attack. Nutrients such as sugars and proteins, and secondary plant compounds such as phenolics, terpenoids, or alkaloids, determine whether or not an insect will feed on a plant and to what extent.

To investigate food preferences by insects or other herbivores, such as deer (Rautio et al. 2008), and the compounds responsible for their choices, these compounds can be added to their diet and tested in feeding bioassays. Here we will add a mixture of phenolic compounds, known as "tannic acid," to the diet of hornworm caterpillars. The tobacco hornworm, *Manduca sexta*, normally feeds on Solanaceae such as tomato or potato plants. For more on the natural history of this insect, consult the information sheet prepared by the biological supply company that ships these caterpillars.

We will perform one of the two bioassays dealing with tannins in insect diet: The compounds to be tested (tannic acid) are *mixed into diet* in varying concentrations. We measure how much chow the caterpillar has consumed and whether the effect is concentration dependent. (The second bioassay – in Chap. 19 – employs the *Leaf disk test*. In this often used bioassay leaf sections of a standard size are treated with the compounds in question.)

Materials Needed

1. Tobacco hornworm caterpillars, from a biological supply house
2. Insect chow, also from a commercial supplier
3. Containers for individual caterpillars

Procedure

Form four working groups.

You will be provided with caterpillars of the tobacco hornworm. They have been food deprived for 2–3 h and should be hungry.

Treated Diet Bioassay

We have prepared four samples of commercial insect chow, each with a different concentration of tannic acid. These samples will be randomly labeled so that the test is double blind (neither you nor the caterpillar will know what is being tested). The concentrations are: 0.05%, 0.5%, and 5% tannic acid. The fourth sample is untreated control.

Data Sheet 18.1 Consumption of TA-treated insect chow over longer time period. Each student group uses one caterpillar

Concentration	Amount of food (grams) at start	Amount left (grams) after feeding	Difference: Amount eaten	Feeding duration (sec)
A				
B				
C				
D				
Total				
Mean				

Each group will test two caterpillars. Give each caterpillar about 1 cm^3 of the treated food, one concentration at a time. We will spread the food in an even layer on paper, marked in millimeter intervals. Observe feeding behavior for 3 min, followed by testing the next concentration. The sequence of concentrations should be different for each student group. One group does ABCD, the second DCBA, the third BADC, and the fourth CDAB, without knowing the concentrations.

Time the duration of feeding; enter times (seconds) during the 3-min period in a data sheet (Data Sheet 18.1).

Results

We will tabulate the following data:

- Plot the time spent on feeding as a function of tannin concentration.
- Test whether differences between feeding times at different concentrations are significant, using Friedman two-way analysis of variance. For comparison of particular pairs of treatments, use the Wilcoxon test.
- Measure or weigh remaining insect chow to determine amount eaten. (Amount eaten can be estimated by counting the millimeter squares under the removed food.) Enter in Data Sheet 18.1.
- Finally, calculate the deterrency index (DI) for each caterpillar

$$DI = (C - T)/(C + T) \times 100,$$

where C is the amount of control diet eaten and T is the amount of treated diet eaten (Data Sheet 18.2).

- Plot the results as bar graph: Treatment is independent variable on x-axis and DI is dependent variable on y-axis.
- Calculate significance of differences between DIs, using an ANOVA. The caterpillars constitute the blocks (replications).

Data Sheet 18.2 Deterrency indices (DI) for both
diets and each caterpillar

	DIs for tannin concentrations		
Caterpillar	0.05%	0.5%	5%
1			
2			
3			
4			
5			
Mean DI			

If more time (in terms of days) is available, study the effect of the tannin diets on
the caterpillars: How do the different tannin concentrations affect weight and survival
of the animals? (see Nomura and Itioka 2002).

References

Nomura M, Itioka T (2002) Effects of synthesized tannin on the growth and survival of a generalist
    herbivorous insect, the common cutworm, *Spodoptera litura* Fabricius (Lepidoptera: Noctuidae)
    Appl Entomol Zool 37:285–289
Rautio P, Kesti K, Bergvall UA, Tuomi J, Leimar O (2008) Spatial scales of foraging in fallow
    deer: Implications for associational effects in plant defences. Acta Oecol 34:12–20

Chapter 19
Leaf Disk Test: Bioassay of Effect of Tannins on Insect Feeding Behavior

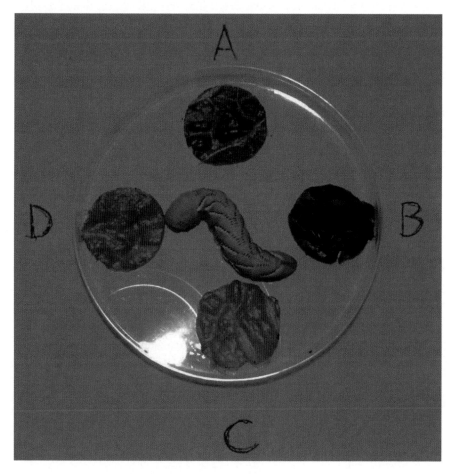

Leaf disk test: A caterpillar of the tobacco hornworm faces a choice of four leaf disks in a Petri dish. The leaf disks are treated with different concentrations of "tannic acid"

D. Müller-Schwarze, *Hands-On Chemical Ecology: Simple Field and Laboratory Exercises,*   105
DOI 10.1007/978-1-4419-0378-5_19, © Springer Science+Business Media, LLC 2009

The feeding behavior of herbivorous insects is guided by plant chemistry. There are specialists and generalists with regard to the range of plant species they attack. Nutrients such as sugars and proteins, and secondary plant compounds such as phenolics, terpenoids, or alkaloids, determine whether or not an insect will feed on a plant and to what extent.

To investigate food preferences by insects and the compounds responsible for their choices, especially reduced consumption, these compounds can be added to their diet and tested in feeding bioassays. Here we will add a commercially available mixture of phenolic compounds (gallic acid and several galloyl glucoses), known as "tannic acid," to the diet of hornworm caterpillars. The tobacco hornworm, *Manduca sexta*, normally feeds on Solanaceae such as tomato or potato plants. For more on the natural history of this insect, consult the information sheet prepared by the biological supply company that ships these caterpillars.

We will perform the *Leaf disk test*, also known as leaf disk assay or leaf disk choice test, the second of two bioassays of tannins in the diet of insects in this book. In this often used bioassay, leaf sections of a standard size are treated with the compound(s) in question. Several circular leaf sections ("leaf disks") (Ali et al. 1999, Filho and Mazzafera 2000, Shields et al. 2008, Wheeler and Isman 2000) or cellulose membrane filters (Hollister and Mullin 1999, Larocque et al. 1999) are presented to a caterpillar in a choice experiment. We measure how much chow the caterpillar has consumed and whether any feeding inhibition is concentration dependent. Regardless of what compounds are being tested, leaf disk tests serve as an important tool in bioassaying feeding inhibitors and stimulants in insects. The cited references are examples of such studies. (In the first tannic acid experiment – Chap. 18 – the tannic acid was *mixed into diet* in varying concentrations.)

Materials Needed

1. Lettuce leaves
2. Tannic acid
3. Petri dishes
4. Tobacco hornworm caterpillars, commercially available

Procedure

To save time, we provide lettuce leaves that are already treated on their surface with different concentrations of tannic acid. The concentrations are 0%, 0.05%, 0.5%, and 5% tannic acid.

Cut out 2.5-cm² large disks from lettuce leaves with a cork borer. Your "cookie cutter" has a diameter of 1.8 cm and makes a 2.55-cm² large hole.

Place four disks – each with a different treatment – evenly spaced into a circle in one Petri dish (see Figure). There should be three concentrations of tannic acid, and one untreated control in each dish.

Each student group works with one caterpillar. Observe feeding behavior for 15 min.

1. Which disk is approached first, second, and so forth?
2. How long does the caterpillar feed on each disk?
3. Record time spent feeding on each disk.
4. Measure areas eaten. In case of irregular pattern, one can weigh the remaining portion of the disk and calculate percent eaten. Or, if class time is very limited, we can estimate the portion of the disk consumed: 1/4, 1/3, etc. and convert this to square millimeters.

For the event that the caterpillars feed very slowly, the instructors have set up some Petri dishes with caterpillars and treated leaf disks beforehand, and you can measure the removed areas of these disks.

Results

With the obtained data, compute the *deterrency index* (DI)

$$DI = (C - T)/(C + T) \times 100,$$

where C is the amount of control eaten, and T is the amount of treatment food eaten (Data Sheets 19.1 and 19.2).

Data Sheet 19.1 Leaf disk test: Results for one caterpillar (Each student group uses one caterpillar)

Concentration	Portion of disk eaten	Portion of control disk eaten	Deterrency index (DI)	Comments
0.05%				
0.5%				
5%				

Data Sheet 19.2 Pooled results from all caterpillars

| Caterpillar | Deterrency index (DI) | | | Remarks |
	0.05%	0.5%	5%	
1				
2				
3				
4				
5				
6				
Total				
Mean				
St. Error				

- Plot the results as bar graph: Treatment is independent variable on x-axis and DI is dependent variable on y-axis.
- Determine significance of differences among the treatments by analysis of variance. The caterpillars are the blocks (replicates) as mentioned in Chap. 18.

References

Some papers that have employed the leaf disk test

Ali MI, Bi JL, Young SY, Felton GW (1999) Do foliar phenolics provide protection to *Heliothis virescens* from a baculovirus? J Chem Ecol 25:2193–2204

Filho OG, Mazzafera P (2000) Caffeine does not protect coffee against the leaf miner *Perileucoptera coffeella*. J Chem Ecol 26:1447–1464

Shields VDC, Smith KP, Arnold NS, Gordon IM, Taharah E, Shaw TE, Waranch D (2008) The effect of varying alkaloid concentrations on the feeding behavior of gypsy moth larvae, *Lymantria dispar* (L.) (Lepidoptera: Lymantriidae). Arthropod–Plant Interact 2:101–107

Wheeler DA, Isman MB (2000) Effect of *Trichilia americana* extract on feeding behavior of Asian armyworm, *Spodoptera litura*. J Chem Ecol 26:2791–2800

Cellulose membrane filter disks have been used in the following studies

Hollister B, Mullin CA (1999) Isolation and identification of primary metabolite feeding stimulants for adult western corn rootworm, *Diabrotica virgifera virgifera* Leconte, from host pollens. J Chem Ecol 25:1263–1280

Larocque N, Vincent C, Belanger A, Bourassa J-P (1999) Effects of tansy essential oil from *Tanacetum vulgare* on biology of oblique-banded leafroller, *Chorisoneura rosaceana*. J Chem Ecol 25:1319–1330

Chapter 20
Two-Way Choice Test for Social Odors in Mice

Two-way choice apparatus (T-maze) for testing mouse responses to odors of conspecifics

D. Müller-Schwarze, *Hands-On Chemical Ecology: Simple Field and Laboratory Exercises,*
DOI 10.1007/978-1-4419-0378-5_20, © Springer Science+Business Media, LLC 2009

This exercise and the next both deal with scent communication in mice. We practice two techniques frequently used in the Animal Behavior laboratory: In this first experiment, we test a mammal's response to conspecific odors in a two-way choice apparatus, also called a Y- or T-maze, an often used bioassay device. [In the following experiment (Chap. 21), we observe and quantify scent marking behavior in response to two different stimuli in an "open field."]

House mice (*Mus musculus* or *M. domesticus*) provide a good model of scent marking in mammals in general. They live in demes, large groups of related individuals. As in many other social mammals, mice mark their territories and home ranges with urine. Both sexes excrete in their urine signaling and priming pheromones that carry a great variety of information. To test what kinds of olfactory signals mice of certain age, sex, and status categories are able to discriminate, we can employ a two-way choice test. In the following, we survey some of the olfactory signals that play important roles in the life of a house mouse.

Urine marks signal individuals' group sex, maturity, group membership, and dominance in an area. In addition, mouse urine also contains important chemical signals that regulate sexual behavior. Some of these signals strongly depend on genetic dispositions. For instance, the Major Histocompatibility Complex codes for signals that affect mate choice: mice choose mates with nonparental urine odors (Yamazaki and Beauchamp 2007). Further, urinary odors vary with hormonal status. Even intrauterine hormonal stimulation of mouse embryos, such as by neighboring male sibling embryos, can androgenize females and change their urinary odors in turn (Vom Saal and Bronson 1980; Vom Saal 1989; Drickamer 2001a, b; Ryan and Vondanbergh 2002).

Dominant adult males mark very frequently. This advertises their aggressive dominance over the other resident and intruder males. Dominant males overmark other males' urine marks, while marks by dominant males are merely investigated by others. Dominant male marks guide dominant and subordinate males to stay within their territory and to avoid areas marked by other dominant males (Hurst 1990a). Juveniles use urine marks to stay within their parental territory.

Urine marking is also important in female–female communication. Resident breeding females countermark breeding female urine, especially of neighbors. Females appear to advertise their dominant breeding status to other females by means of urine marks (Hurst 1990b).

Urine marks also serve communication between the sexes. Dominant males countermark female urine marks at a high rate. Females are attracted to marks by their resident dominant male, but avoid those from neighbor and unfamiliar dominant males (Hurst 1990c). These are just some examples of odors that mice encounter in their daily life and use to extract vital information that guides their behavior.

In the T-maze we can test the responses of a male or female to a urine mark of the opposite sex, or from dominant or subordinate individuals of the same sex, or some other difference of interest. Among the many studies using two-way choice apparatuses for mice, a good example is a paper by Krasnov and Khokhlova (1996) that deals with mice responding to odors of another rodent species. We will test responses of males and females to urine of the same and different sex in a *two-way choice apparatus (T-Maze)*.

Materials Needed

We will work in four groups, each with one T-maze. Materials needed by each group are as follows:

1. T-maze
2. One male and one female mouse
3. A piece of filter paper each marked by a male or female mouse with urine
4. A data sheet

Procedure

We will test responses of male and female mice to urine from males and females. Each group will first run a male, then replace the tubes with clean ones, and run a female. To make the test more specific, we can first determine the dominance status of the odor donor by staging encounters, and then specifically observe responses to dominant or subordinate individuals.

Place a filter paper with the *urine of a male* mouse at the end of one arm of the T-maze and a clean filter paper in the other arm. In a second version, juxtapose male and female urine odor in the two arms.

Place a *male* mouse at the start of the maze and close the cap.

Watch the mouse's behavior for 3 min.

Record:

1. *Latency*: The time it takes until the mouse moves forward in the maze.
2. *First choice*: Male or female odor.
3. *Time spent in left or right arm of maze.*

Exchange the soiled arms for clean ones.

Use a new *male-soiled* filter paper. Run a *female* mouse.

Repeat running first a *male*, then a *female* mouse responding to *female* urine on filter paper either pained with a clean filter paper or male urine.

The class will compile mean values obtained from the four mazes run simultaneously by the four groups.

Results

- Do males and females differ in their behavior?
- Do male and female urine release different responses?
- Tabulate and graph data. Use Data Sheet 20%.

Data Sheet 20.1 Two-way choices of odors by mice in T-maze

T-maze				
Group	Latency	Time near male odor	Time near female odor	First approach to F or M odor
Responses by male				
1				
2				
3				
4				
Totals				
Mean				
Responses by female				
1				
2				
3				
4				
Totals				
Mean				

- Test the data for significance in two analyses: First, concentrate on the proportion of time a sniffer spends with the male odor. Use a two-sample t test for proportion of time the male spends on the ♂ side, and the proportion of the time the female spends on the ♂ side.
- Next, concentrate on the attractiveness of the odors: If equally attractive, the animals would spend a proportion of 0.5 on each side. In a one-sample t test, compare the measured proportion of time against that mean of 0.5.
- Write report.

References

Drickamer LC (2001a) Intrauterine position effects on rodent urinary chemosignals. In: Marchlewska-Koj A, Lepri JJ, Müller-Schwarze D (eds) Chemical signals in vertebrates, vol 9. Kluwer, New York, NY, pp 211–216

Drickamer LC (2001b) Ecological aspects of house mouse urinary chemosignals. In: Marchlewska-Koj A, Lepri JJ, Müller-Schwarze D (eds) Chemical signals in vertebrates, vol 9. Kluwer, New York, NY, pp 35–41

Hurst JL (1990a) Urine marking in populations of wild house mice Mus domesticus Rutty. I. Communication between males. Anim Behav 40:209–222

Hurst JL (1990b) Urine marking in populations of wild house mice Mus domesticus Rutty. II. Communication between females. Anim Behav 40:223–232

Hurst JL (1990c) Urine marking in populations of wild house mice Mus domesticus Rutty. III. Communication between the sexes. Anim Behav 40:223–243

Krasnov B, Khokhlova I (1996) Discrimination of midday jird's odour by house mice. Anim Behav 52:659–665

Ryan BC, Vandenbergh JG (2002) Intrauterine position effects. Neurosci Biobehav Rev 26:665–678

Vom Saal FS, Bronson FH (1980) Sexual characteristics of adult female mice are correlated with their blood testosterone levels during prenatal development. Science 208:597–599

Vom Saal FS (1989) The production of and sensitivity to cues that delay puberty and prolong subsequent oestrous cycles in female mice are influenced by prior intrauterine position. J Reprod Fert 86:457–471

Yamazaki K, Beauchamp GK (2007) Genetic basis for MHC-dependent mate choice. Adv Genet 59:129–145

Chapter 21
Scent Marking in Mice: Open Field Test

Mouse in open-field test

D. Müller-Schwarze, *Hands-On Chemical Ecology: Simple Field and Laboratory Exercises,*   115
DOI 10.1007/ 978-1-4419-0378-5_21, © Springer Science+Business Media, LLC 2009

This exercise is the second of the two experiments dealing with communication via scent marks in mice. In this *Open-field Test*, we observe scent marking behavior of male mice in response to presence or absence of urine stimuli in their environment. (In fact, the "open field" arena is enclosed and covered.)

Where mice live undisturbed for some time, their urine marks accumulate and are spread over most surfaces. At topographical edges such as walls, pipes, or rafters, posts of concentrated urine pile up over time. Dominant adult males mark very frequently. This advertises their aggressive dominance over other resident and intruder males. Dominant males overmark other males' urine marks, while marks by dominant males are merely investigated by others. Dominant male marks guide dominant and subordinate males to stay within their territory and to avoid areas marked by other dominant males (Hurst 1990a). Juveniles use urine marks to stay within their parental territory.

Dominant males mark differently from subordinate males: they cover an entire area, while subordinates urine-mark along the walls of their enclosure or cage, and in fewer and larger patches. These patterns can be visualized under ultraviolet light (Desjardins et al. 1973).

Urine marking is also important in female-female communication. Resident breeding females countermark breeding female urine, especially of neighbors. Females appear to advertise their dominant breeding status to other females by means of urine marks (Hurst 1990b).

Finally, urine marks also serve communication between the sexes. Dominant males countermark female urine marks at a high rate. Females are attracted to marks by their resident dominant male, but avoid those from neighbor and unfamiliar dominant males.

Here, in a sequence of tests, we observe the behavior of a male or a female as it moves around an arena with urine-marks by a former occupant. First, only one floor tile of many will carry urine marks, while the others are clean. After that, we present the male or female with a thoroughly marked arena that has one clean tile inserted. We will see that "no odor" (or edge of marked area) is a powerful stimulus that triggers specific behaviors.

Materials Needed

1. Open-field arena with lid
2. Clean tiles for arena floor
3. Four extra clean tiles (or clean after first test)
4. Two tiles soiled by males
5. Two tiles soiled by females
6. Two male and two female mice

Use an open-field arena that has 8 × 8 square tiles. Hurst (1988, 1989) used an arena 1.2 × 1.2 m large with 15 × 15 cm tiles. (One commercially available arena has 7 × 7 tiles, but will work, too). We will divide the arena into four quadrants (these may not be of entirely equal size; see above). We will run two tests.

Test 1a: Sex Differences

Response of male and female mice to a mainly *clean arena that has one tile inserted which carries urine marks from either a male or a female mouse.*

Quadrant 1 has *male urine* on this one tile, and a *male* will be placed into the arena.

Quadrant 2 has *male urine*, and a *female* will be placed into the arena.

Quadrant 3 has *female urine*, and a *male* will be placed into the arena.

Quadrant 4 has *female urine*, and a *female* will be placed into the arena.

Observe the behavior of each animal toward the clean tiles and toward the soiled tile. This part requires good observation and description of what the animal is doing. Describe what you see: walking, running, stopping, turning, sniffing, urinating, defecating, etc.

How does the animal change its behavior as it encounters the soiled tile?

Test 1b: Effect of Age of Urine

Male mice mark competitively mostly to *fresh* urine of other males when compared to aged urine. We can test this by placing fresh urine, 24-h old urine, and 7-days old male urine marks (covered with some wire mesh) in the arena. Observe the marking behavior directly, and count the number of marks present after about 10 h (or earlier). Note that these different types of urine can interact in their effects: Even though 7-day-old urine is of less interest, it will be marked more, if also fresh urine is present at the same time. But the 7-day-old urine will be marked little if it is paired with 24-h old urine (Humphries et al. 1999).

Test 1c: Scent Marking by Different Types of Males

Mouse embryos of both sexes are hormonally affected by their male neighbors in utero. For males, depending on whether no, one, or two males reside next a particular male in utero, he is designated 0M, 1M, or 2M. These males differ anatomically in their anogenital distance (AGD). This AGD can be measured in mice from 10 days of age on. (AGDs of males in our experiments ranged from 10–14 mm). Thus, measuring AGD is an indirect measure of male hormone effects during their development. The AGD of females is also variable, depending on male neighbors in utero. (We worked with female mice, whose AGDs varied between 5 and 9 mm.)

Measure the AGD of several males. Run the most extreme individuals along the AGD continuum in the open field and record their scent marking.

Test 1d: Responses to Urines from Different Types of Males

Use urines from males that differ in their AGD as stimuli and record the responses of other males. Do you find fresh urine from a male with a large AGD more potent than that from a male with a shorter AGD? What does it tell us about prenatal effects on competitiveness in later life?

Test 2: Effect of Absence of Scent Marks

Response of male or female to *soiled arena* that has inserted *one clean tile*. We will use the arena as the previous animals have left it, but now replace one of the soiled tiles with a clean one.

Results

- Observe and describe what a male, then a female does when it encounters the "odd" tile. Does it run, stop, turn, sniff, urinate, etc.?
- Enter the observation in your data sheet (Data Sheet 21.1).
- Based on your own and/or the data provided below, compare the observed number of visits to the single tile with the expected number of visits if the mouse had moved around randomly in the open field. Use a binomial test.

Data Sheet 21.1 Behavior of mice in open field arena

Female sniffs tile with female urine					
	Clean arena		Soiled arena		Remarks
	Soiled tile	Rest of clean arena	Clean tile	Rest of soiled arena	
Stop/sniff					
Expected frequency					
Fecal pellets					
Urinate					
Male sniffs tile with male urine					
Stop/sniff					
Expected frequency					
Fecal pellets					
Urinate					

Conclusions

Try to answer the following questions:

1. Is the single, "unusual" tile visited significantly more often than expected if all tiles were visited equally?
2. Is there a significant difference between behavior toward the clean tile in a soiled arena and the soiled tile in a clean arena?
3. Do male and female differ in their behavior?
4. Do you find the greatest sex difference in behavior toward the clean or soiled tile?
5. What does deposition of fecal pellets indicate?
6. Do the sexes differ in their urine marking?
7. What do the findings mean in the context of the social organization of house mice?

Some Previous Results

Of 12 tiles, one was the "odd one" (a tile soiled by a female or a male in a clean arena; or a clean tile in an arena soiled by a female or a male). A mouse was observed to stop and sniff at the following frequencies:

Open field test: Frequency of stopping and sniffing a tile				
	Clean arena		Soiled arena	
	Soiled tile	All other tiles (clean)	Clean tile	All other tiles (soiled)
♀ sniffs ♀ tile	21	28	15	21
♂ sniffs ♂ tile	21	59	7	40

Compare these results with your own data. Try to explain any differences between these and your results. If needed, use these frequencies in lieu of your own data. Use a binomial test for the difference between the observed proportion and an assumed proportion (here 1/12 of all visits by the mouse, i.e., the share of 1 tile out of 12, had the mouse moved around randomly).

References

Desjardins C, Maruniak JA, Bronson FH (1973) Social rank in house mice: Differentiation revealed by ultraviolet visualization of urinary marking patterns. Science 182:939–941

Humphries RE, Robertson DHL, Beynon RJ, Hurst JL (1999) Unraveling the chemical basis of competitive scent marking in house mice. Anim Behav 58:1177–1190

Hurst JL (1990a) Urine marking in populations of wild house mice *Mus domesticus* Rutty. I. Communication between males. Anim Behav 40:209–222

Hurst JL (1990b) Urine marking in populations of wild house mice *Mus domesticus* Rutty. II. Communication between females. Anim Behav 40:223–232

Hurst JL, Robertson DHL, Tolladay U, Beynon RJ (1998) Proteins in urine scent marks of male house mice extend the longevity of olfactory signals. Anim Behav 55:1289–1297

Chapter 22
Human Body Odor Discrimination: T-Shirt Experiment

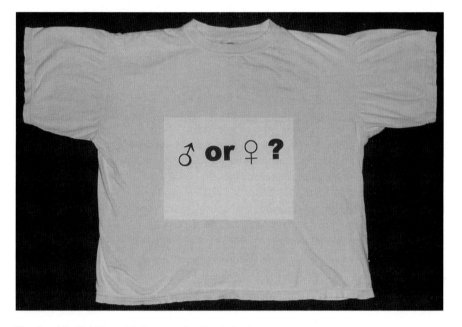

Simple white T-shirt used in human odor discrimination test

D. Müller-Schwarze, *Hands-On Chemical Ecology: Simple Field and Laboratory Exercises,*   121
DOI 10.1007/ 978-1-4419-0378-5_22, © Springer Science+Business Media, LLC 2009

Body odors play a more important role in human behavior than we realize. Mothers can identify their infant's odor when presented with several garments worn by their own and other children. Breast-fed infants learn to recognize the axillary odor of their mothers, while bottle-fed infants do not (Cernoch and Porter 1985). Children between the ages 3 and 5 years correctly discriminate their mother's T-shirt from those of other women (Montagner 1974). Adults can identify the gender of the wearer by smelling a worn T-shirt (Russell 1983; Schleidt et al. 1981). People can even identify their own odor on a T-shirt among several shirts. In general, female odors are perceived as more pleasant than male odors. But pleasantness is correlated with the strength of the odor, so that a strong female odor can be misclassified as male (Doty et al. 1978).

Can humans reliably identify gender and individuals by body odors? The answer to this question matters very much in a number of contexts of social and sexual behavior such as parent–offspring bonding, mate choice, or nepotism. A classical experiment has examined the role of the *axillary odor*, the most powerful human scent. This odor is the result of bacterial action on the secretion of the axillary glands. These skin glands are of the *apocrine* type which is a modified sweat gland that produces an aqueous secretion.

In this experiment, a person wears a T-shirt for 24 h. After that time, the shirt is placed in a bucket or paper bag, so that it can be smelled, but not seen. In addition to discrimination between male and female shirts, the participants are classified into the male or female category, to determine whether the genders differ in discrimination acuity. Further, subjects may differ in their degree of familiarity or relatedness with the odor donor(s). We can then ask whether subjects are able to pick out individual odors of close genetic relatives, spouses, or close friends from among a series of odors of strangers.

The experiment at hand focuses on gender discrimination.

Materials Needed

Per student:

1. One T-shirt
2. One paper bag
3. One index card
4. Latex gloves for laboratory assistant
5. Identical paper bags (number depends on class size)

Procedure

The schedule described here applies to a Tuesday class.

Monday

Wear the T-shirt continuously for 24 h (Monday morning to Tuesday morning). The shirt should not be worn during heavy exercise. Do not use deodorants or odorous substances (perfume, cologne) immediately before or during the wearing period. Do not shower with perfumed shampoo or soap on Monday morning. Instead, take a normal shower on Sunday night. But on Monday morning, just rinse off with water. When the wearing period is completed on Tuesday morning, place your T-shirt in the bag and write your name on the 3×5 card and put it in the bag.

Tuesday

Bring the bag with the worn T-shirt to class. Immediately before the test, the laboratory assistant will transfer the T-shirts to different numbered, identical (paper) bags. The experimenter will make and keep a number–name key.

Numbers of male and female students will not be equal in most classes. To eliminate guessing on the basis of the more frequent gender represented by the shirts, trim the number of shirts to be sniffed to equal numbers, say, ten male and ten female shirts. (Or, if the laboratory assistant really wants to be sneaky, use only shirts of one gender and ask to which gender they belong. Such research has been done.)

Each student will be provided with a score sheet upon which he/she will try to determine for each shirt whether it was worn by a female or male, and whether it was the subject's own shirt (Data Sheet 22.1).

Subjects can sniff shirts (in bags, of course) one at a time without touching them. Unlimited exchange and resniffing is allowed until everyone is satisfied that he/she has scored each shirt to the limit of his/her ability to do so.

Results

- Score sheets are marked for accuracy.
- The results will be scored for departure from random choice, first for identifying "self": If each student randomly labeled 20 shirts as "self," he/she would be correct in 1 out of 20 choices. Use a binomial test for the difference between an observed and an assumed proportion (here 1/20) for each student separately.
- For ability to discriminate gender by odor, use one subset, e.g., only five male and five female shirts. In a binomial test, compare one person's proportion of labeling shirts as "male" with an assumed random proportion of 0.5.

Data Sheet 22.1   One student's assessment of the odors of T-shirts of the entire class

Date:				Name:			
Shirt #	Male	Female	Self	Shirt #	Male	Female	Self
1				12			
2				13			
3				14			
4				15			
5				16			
6				17			
7				18			
8				19			
9				21			
10				22			
11				Total correct			

References

Cernoch JM, Porter RH (1985) Recognition of maternal axillary odor by infants. Child Dev 56:1593–1598

Doty RL, Orndorff MM, Leyden J, Klingman A (1978) Communication of gender from human axillary odors: Relationship to perceived intensity and hedonicity. Behav Biol 23:373–380

Montagner H (1974) Communication non verbale et discrimination olfactive chez le jeune enfant: Approache éthologique. In: Morin E, Piatelli-Palmarini M (ed) L'Unite de l'homme. Seuil, Paris

Porter RH, Moore JD (1981) Human kin recognition by olfactory cues. Physiol Behav 27:493–495

Russell MJ (1983) Human olfactory communications. In: Müller-Schwarze D, Silverstein RM (eds) Chemical Signals in Vertebrates, vol 3. Plenum, New York, NY

Schleidt M, Hold B, Attili G (1981) A cross-cultural study on the attitude towards personal odors. J Chem Ecol 7:19–31

Chapter 23
Coping with Plant Volatiles in Spicy Food ("Burping Exercise")

Spicy foods contain Plant Secondary Metabolites that humans discharge by eructation (gas belching, burping) as one of the body's first detoxication responses to xenobiotics

D. Müller-Schwarze, *Hands-On Chemical Ecology: Simple Field and Laboratory Exercises,*   125
DOI 10.1007/ 978-1-4419-0378-5_23, © Springer Science+Business Media, LLC 2009

Humans, like many other land vertebrates, take up a large variety of plant secondary metabolites (PSMs) and other xenobiotics in their diet. Xenobiotics constitute a wide category of compounds, including synthetic ones, in the environment that are potentially harmful when ingested, especially in large amounts. Mammals and birds have evolved many mechanisms to render harmless these PSMs and other toxins (Iason 2005). These detoxication mechanisms include elimination of the compounds themselves, metabolizing them into less toxic and/or better excretable compounds, or developing tolerance, as for instance, garter snakes have in response to the very potent tetrodotoxin of newts (Brodie et al. 2002). These countermeasures start in the mouth where plant volatiles escape during mastication. Normal body temperature suffices to volatilize monoterpenoids (Welch et al. 1989). Therefore, in mammals, such plant constituents in the diet can be reduced by mastication and eructation, among other processes such as absorption and excretion (Welch et al. 1989).

Once in the stomach, PSMs still can be belched up, in addition to swallowed air or other gases such as carbon dioxide. Most of us have experienced this normal physiological reaction after eating foods such as cucumbers, hot peppers, or garlic, or drinking coffee or herbal teas. If obserbed through the gut, metabolic deactivation takes place in the liver by the hepatic microsomal cytochrome P450. There, the PSMs are activated by oxidation, reduction, or other processes and then conjugated with glucuronic acid or other compounds to make them excretable in urine or bile.

This exercise addresses the question: Is burping (belching) part of a detoxication mechanism? Does the upper GI tract function in ridding the body of potentially toxic plant compounds? Specifically, does burping serve in removing plant volatiles from ingested spicy food?

Procedure

At routine dinners, keep record of spicy food eaten. Classify into "bland," "somewhat spicy," and "very spicy." Record numbers of burps within 2 hours after the meal for 3 days (1 day for each spiciness class) in a table as given in Data Sheet 23.1.

Data Sheet 23.1 Eructation after spicy meals

Date	Dish eaten	Level of spiciness	Number of burps within 2 h	Remarks

Results

- In class, we will combine the data from different students, and compute a score (Number of burps per episode):

Total for all students	(Almost) no spice ("bland")	Somewhat spicy	Very spicy food
Number of episodes			
Burping frequency			
Rate of burping (mean # burps/episode)			

- To test for statistical significance, use a randomized ANOVA. Each student is a block. Follow this with Tukey's test.
- Finally, we will graph the results.

What is the Conclusion?

_____

_____

_____

_____

_____

Are the results consistent with the hypothesis that burping aids in detoxing the organism by removing plant volatiles from the upper GI tract?

Discuss sources of error in this opportunistic exercise: The pitfalls of self-observation; other sources of burping such a swallowed air when eating fast; and carbonated beverages. Consider also individually different perceptions of what is "spicy" or "very spicy." What about amounts eaten? In a really controlled experiment, these points should be addressed. If all ate the same amount of the same dish, how would individual differences affect responses to this "standard diet"?

Some Previous Results

The mean number of burps per meal for students in four different courses was as follows. Compare these results with yours.

After bland meal	After slightly spicy meal	After very spicy meal
1.3	1.76	2.4
2.13	2.71	4.29
1.65	1.76	4.12
1.0	2.3	3.83

References

Brodie ED, Ridenhour BJ, Brodie ED (2002). The evolutionary response of predators to dangerous prey: Hotspots and coldspots in the geographic mosaic of coevolution between garter snakes and newts. Evolution 56(10):2067–2082.

Iason G (2005) The role of plant secondary metabolites in mammalian herbivory: Ecological perspectives. Proc Nutr Soc 64:123–131. doi:10.1079/PNS2004415

Welch BL, Pederson JC, Rodriguez RL (1989) Monoterpenoid content of sage grouse ingesta. J Chem Ecol 15:961–969

Section III
Perspectives of Experiments: Development and Application

Chapter 24
Miscellaneous Experiments Chosen by Students

Over the years, our students have chosen a great variety of independent projects in the Chemical Ecology courses. Some such as those with dogs, cats, or mice can be done in most places, while others took advantage of special facilities (e.g. fish hatchery) or animal species available to a particular student (e.g. captive reptiles). All studies stimulated discussion and spawned ideas for further research, whether they were basic or applied, field studies, or indoor projects. Here are some of the topics that were successfully completed with great enthusiasm in more recent courses. It should be noted that the courses focused on vertebrates.

- Responses of male red-spotted newts (*Notophthalmus viridescens*) to water-borne female odors
- Feeding ecology of grouse and phenolics in their food plants
- Feeding responses of domestic rabbits to plant secondary metabolites (Capsaicin, caffeine)
- Reasons for proliferation of poisonous weeds in livestock pastures
- Effect of tannin and tea grounds on feeding by gray squirrels
- Gustatory responses of lake sturgeon (*Acipenser fulvescens*) juveniles to extracts of natural prey and amino acids
- Tongue flicking behavior of snakes in response to odors from small mammals
- Effects of bait type and location on trapping success with deer mice
- Food choices of Norway rats
- Tadpole behavior in the presence of predatory fish
- Prey fish (pumpkinseed and common shiner) responses to odors from predatory fish (rock and smallmouth bass)
- Chemical alarm responses in frog tadpoles
- Responses of rodents to cat urine
- Responses of rats and mice to predatory snake odors
- Comparison of effects of commercial squirrel repellents with fox urine
- Oil of Citronella as repellent for dogs and cats
- Camphor as repellent for cats
- Predator odors as squirrel repellents
- Naturally occurring contraceptive compounds in plants
- Attractiveness rating of photos of men and women with and without fragrance present

D. Müller-Schwarze, *Hands-On Chemical Ecology: Simple Field and Laboratory Exercises*,   131
DOI 10.1007/ 978-1-4419-0378-5_24, © Springer Science+Business Media, LLC 2009

Chapter 25
Further Possible Experiments

This final chapter suggests some additional ideas for more exercises that instructors can flesh out and develop into different directions. The cited references are meant to be mere starting points to the literature.

Intraspecific Communication: Pheromones

Response of Female Nematodes to the Male Pheromone

Under a microscope, expose male nematodes to a drop of vanillic acid. Vanillic acid is the sex pheromone of females. The particular species studied was the soybean cyst nematode, *Heterodera glycines*. The male nematode should respond by moving to the odor source, then coiling (Chasnov et al. 2007, Jonz et al. 2001; Meyer and Huettel 1996). You can test concentration effects; compare responses to homologs and analogs; and test for specificity by using pheromones of other nematode species. This experiment is more suited for campuses where nematode research is already going on and animals and pheromones are more readily available.

References

Chasnov JR, So WK, Chan CM, Chow KL (2007) The species, sex, and stage specificity of a *Caenorhabditis* sex pheromone. Proc Natl Acad Sci USA 104:6730–6735

Jonz MG, Riga E, Mercier AJ, Potter JW (2001) Partial isolation of a water soluble pheromone from the sugar beet cyst nematode, *Heterodera schachtii*, using a novel bioassay. Nematology 3:55–64

Meyer SLF, Huettel RN (1996) Application of a sex pheromone, pheromone analogs, and *Verticillium lecanii* for management of *Heterodera glycines*. J Nematol 28:36–42

Social Communication in EarthWorms

When alarmed, earthworms secrete coelomic fluid through dorsal pores, located in the grooves between the segments. To trigger release of the alarm secretion, one can shock an earthworm electrically (obviously not exactly a natural stimulus) with current from two size D batteries while on a wax paper. The worm will secrete coelomic fluid. A 2–cm^2 piece of this paper with secretion can be cut out and presented to another, naive earthworm. Observe the response: rearing up and withdrawing. Compare this behavior with the response to control stimuli such as salt solution or the "normal" mucus from an undisturbed earthworm (Rosenkoetter and Boice 1975).

Earthworms can be purchased from a bait dealer. Use *Eisenia foetida* or *Diplocardia riparia*. (*Lumbricus terrestris* responds positively to coelomic fluid, a behavior which is harder to recognize.) Keep the worms in a bucket in peat moss, old leaves, or commercial earthworm bedding. Worms must be healthy and "calm," i.e., crawling slowly forward, do not jerk about rapidly, and do not tend to crawl backwards or defecate.

References

Rosenkoetter JS, Boice R (1975) Chemical communication in earthworms. In: Price EO, Stokes AW (eds) Animal behavior in laboratory and field. Freeman, San Francisco, CA

Trail Following by Slugs

Terrestrial slugs follow mucus trails back to their daytime resting areas, to find food, or to mate. They follow their own, or each other's trails. Slugs can determine the direction of a mucus trail. Predators, such as snakes or aquatic sciomyzid fly larvae, follow the mucus trail to find their prey. Laying the mucus trail is energetically expensive (Cook 1979, 1992; Davies and Blackwell 2007).

Such chemically marked trails can be altered experimentally: interrupted, erased, or covered with other chemicals. Watch the behavior of the slugs. Is it possible to divert them by laying deviating mucus trails? At what angle can an experimental trail divert a slug from its own? Test direction finding by starting a slug perpendicular to a fresh slug trail. How old can a mucus trail be before it loses its activity? Does a slug follow another species' trail?

References

Cook A (1979) Homing by the slug *Limax pseudoflavus*. Anim Behav 27:545–552
Cook A (1992) The function of trail following in the pulmonate slug, *Limax pseudoflavus*. Anim Behav 43:813–821
Davies MS, Blackwell J (2007) Energy saving through trail following in a marine snail. Proc Biol Sci 274:1233–1236

Pheromone Responses of Insects

Insects communicate between the two sexes by chemical attractants and stimulants. Numerous pheromones have been identified in a variety of insect species, and the literature on insect pheromones is vast. We know particularly well pheromone communication in commercially important species such as bark beetles, and moths that are agricultural pests. Six possible types of pheromone exercises follow.

1. While attraction by pheromones appears to be a straightforward behavior when presented in textbooks, actual experiments can turn out to be challenges. Particularly within the time constraints of a laboratory session of a college course, experiments may not always work out satisfactorily. It is recommended to schedule extra time for such exercises, or, better yet, join some *ongoing pheromone field research*. There may be pheromone traps to be prepared and placed in the field, traps to be checked, or insects to be counted.
2. Another avenue suggests itself: buy *commercial pheromone traps* such as for example those for Japanese beetles, a garden pest, and test the efficacy of such traps. In this case, the downside can be examined: Does the trap attract actually more beetles into the garden?
3. Like many other insects, moths attract mates by long-distance pheromones. Females produce these pheromones in specialized abdominal glands. Chemically, they are acetates, often active in precise mixtures of geometric isomers. Males fly upwind, following the females pheromone plume to the source, and mating ensues. In a typical experiment, a female moth, or just the pheromone, serves as odor source. An air current from that source helps to attract males who fly upwind to the pheromone source and attempt to mate. With this technique, we can compare the effects of known pheromones from different, related species on one species (*species specificity*). We can also test the attractiveness of different compounds that are structurally similar to a known pheromone. In the laboratory, a *wind tunnel*, where available, is ideal, for this experience.
4. Since most schools do not have a wind tunnel readily available, a small one can be improvised by using a fan that blows air across some tabletop enclosure made out of plywood, plastic, plexiglass, or cardboard. The cover has to be transparent for observation.
5. We have also tried to attach filter papers with pheromone, control odors, and no odor, respectively, in the four corners of the laboratory or classroom near the ceiling. Moths can be released one by one in the center of the room, and their flight pattern monitored and scored. Dr. Stephen Teale initiated this experiment.
6. Trail following:
   - *Trail following in caterpillars.* Tent caterpillars (example: Eastern Tent Caterpillars in North America) venture out from their shelters, the "tents," in a tree to consume leaves. They follow chemical trails, just as ants and termites do. These odor trails lead along branches and twigs. Outbound foragers follow the trails of successful returning foragers. Among such "recruitment trails," caterpillars prefer fresh trails and trails laid by fed caterpillars.

Where trees infested with tent caterpillars are available, they can be studied as they follow odor trails along branches or twigs of trees. Their odor trails along branches can be experimentally manipulated by interrupting, reversing, or covering them with other odors (Travis 2003). Look for apple trees or other members of the Rosaceae family in spring.

- *Trail pheromones of termites* can be mimicked by traces of ballpoint pen ink (Chen et al. 1998). Blue Papermate ball point pens work best. The "scent trail" must be fresh. Provide such ballpoint pen ink trails that are just made, and 30, 60, 90 and 120 s old.

  At what age of the trail do the termites ignore the trail?

  What does it tell you about the function of chemical trails in the lives of the termites?

- *Trail pheromones of ants* offer many opportunities for experimentation. For example, the Argentine ant's trail pheromone is (Z)-9-hexadecanal. Trail following can be disturbed by an overdose of pheromone. The investigators (Suckling et al. 2008) used the pheromone on wax-coated sand (1 g sand, 0.2 g wax, 25 mg pheromone per m^2).

References

Chen J, Henderson G, Laine RA (1998) Isolation and identification of a 2-phenoxyethanol from a ballpoint pen ink as a trail-following substance of *Coptotermes formosanus* Shiraki and *Reticulitermes* sp. J Entomol Sci 33:97–105

Suckling DM, Peck RW, Manning LM, Stringer LD, Cappadonna J, El-Sayed AM (2008) Pheromone disruption of argentine ant trail integrity. J Chem Ecol 34:1602–1609

Travis H (2003) Pheromone caterpillar trails: An easy lab exercise for the classroom. Am Biol Teacher 65:456–461

Visualization of Pheromone Pulses in Goldfish Urine with Isosulfan Blue

Female Goldfish release a spawning pheromone in their urine. They release urine in pulses every 2–4 min. These pulses can be made visible by a blue dye in the urine (Appelt and Sorensen 1999).

Inject a 13–24 g goldfish with isosulfan blue (this substance is used for lymphography in humans). The dye is available from Sigma Chem. Co in St. Louis, MO, as "patent blue violet, 'purified'." Mix with buffered saline solution. Inject this blue dye (180 µg/µl solution). Use 60 µg isosulfan blue per gram body weight. Observe urine pulses against a white background (styrofoam, white gravel).

References

Appelt CW, Sorensen PW (1999) Freshwater fish release urinary pheromones in a pulsatile manner. In: Johnston RE, Müller-Schwarze D, Sorensen PW (eds) Advances in chemical signals in vertebrates. Kluwer, New York, NY, p 247

Mate Choice by Salamanders in Y-Maze

To test their ability to discriminate individuals, male and female salamanders (in the US: several species of the genus *Plethodon*) are given a choice of odors in a Y-maze. The odors tested can be male vs. female; mate vs. strange individual of opposite sex; individuals from different populations or closely related species (Dawley 1985).

References

Dawley EM (1985) Evolution of chemical signals as a premating isolation mechanism in a complex of terrestrial salamanders. In: Duvall D, Müller-Schwarze D, Silverstein RM (eds) Chemical signals in vertebrates, vol. IV. Plenum, New York, NY, p 221

Social and Hormonal Influences on Scent Marking in Gerbils and Hamsters

This experiment deals with scent marking related to skin gland size and resembles Chaps. 20 and Chaps. 21 in the main text (Scent marking in mice). Male gerbils scent mark with their ventral glands. Levels of testosterone determine the size of the ventral gland (Thiessen et al. 1968). Therefore, measuring the gland size indicates the testosterone-correlated status of a male gerbil and can be used as independent variable when counting scent marking frequency. Second, marking in response to scent marks by another male gerbil can be quantified. Also, we can determine how the presence of other gerbils in the area affects scent marking (Drickamer 1975).

References

Drickamer L (1975) Hormonal and social influences on the scent-marking behavior of the Mongolian gerbil. pp. 83–85 In: Price EO, Stokes AW (eds) Animal behavior in laboratory and field. (Price EO, Stokes AW, eds), Freeman, San Francisco, CA, pp 83–85
Thiessen DD, Friend M, Lindzey G (1968) Androgen control of territorial marking in the Mongolian gerbil (*Meriones unguiculatus*). Science 160:432–442

Visualization of Mouse Scent Marks

Urine marks of mice in a laboratory setting can be made visible by spraying them with ninhydrine. The numbers of marks, or area marked, are measured by laying a grid over the marked arena (Roberts and Gosling 2004). Second, urine marks fluoresces under ultraviolet light (Desjardins et al. 1973).

References

Desjardins C, Maruniak JA, Bronson FH (1973) Social rank in house mice: Differentiation revealed by ultraviolet visualization of urinary marking patterns. Science 182:939–941
Roberts SC, Gosling LM (2004) Manipulation of olfactory signalling and mate choice for conservation breeding: A case study of the harvest mouse. Conserv Biol 18:548–556

Method to Remove Odor-Carrying Lipids from Body Surface of Small Mammals

To observe the role of chemical cues in the sexual behavior of males of small mammals, we can alter the odor of females. Hair is a great substrate for odor-carrying lipids. For example, the behaviorally active body odors of sheep reside in the wool fat, also known as "suint." Removing lipids from the fur of female mice reduces the lipophilic odorants significantly. Hair near skin glands is often anatomically modified into *osmetrichia* to hold sebum on its cuticular surface (Müller-Schwarze et al. 1977).

Observe the behavior of a male mouse toward an untreated female for 5 min. Record behaviors such as approach, nasonasal sniffing, nasogenital sniffing, mounting, and chasing. Enter in a data sheet.

Dissolve sodium lauryl sulfate, a surfactant with a detergent's amphiphilic properties, in water. (This compound creates lather in car washes, shampoos, bubble baths, etc.). We use it here because it does not have an odor of its own. Use 750-ml powder in 15 l of warm water (or prepare a smaller volume).

Let a female mouse or rat swim in the solution for 5 min. This removes 50% of the hair fat (Johnston 1986). Rinse the animal in warm water for 1.5 min. Dry it with a paper towel. Observe the sexual response of a male and compare it to his behavior vis-a-vis an intact female.

Observe the behavior of the same male toward the same female after treatment. Compare "before" and "after" behavior of the male.

References

Johnston RE (1986) Effect of odors on male sexual behavior. Behav Neural Biol 46:168–188
Müller-Schwarze D, Volkman NJ, Zemanek K (1977) Osmetrichia: Specialized scent hairs in black-tailed deer. J Ultrastruct Res 59:223–230

Interspecific Responses: Allomones (Emission of Chemicals that Benefit the Sender)

Chemical Defense in Insects: Ladybird Beetles, Stick Insects, Stinkbugs, or Other Insects

Insects offer endless opportunities for experiments with defense secretions. For a review, see Eisner et al. 2005. Upon gentle contact with a forceps insects will release defense secretions. Keep the insect that just has released the defense secretion in a vial with a wire screen cover, or in a tiny wire cage of the type honey bee queens are shipped in through the mail. Observe responses by other insects or predators such as spiders, lizards, frogs, birds, or small rodents.

Second, test the effect of the defense secretions alone, in the absence of its donor. For rodents, a circular arena has been used. The test odors were introduced through holes in the wall of the arena. A rat's visits to holes with defense odor were then compared with visits to holes with cheese odor or no odor (Bouchard et al. 1997).

Do you find your test animal avoids the defense secretion? For how long? How do interpret this?

References

Bouchard P, Hsiung C-C, Yaylayan VA (1997) Chemical analysis of defense secretions of *Sipyloides sipylus* and their potential use as repellents against rats. J Chem Ecol 23:2049–2057
Eisner T, Eisner M, Siegler M (2005) Secret weapons: Defenses of insects, spiders, scorpions and other many-legged creatures. Belknap Press, Cambridge, MA

Plant Defense: Preformed and Induced Resin Defense by Pine Trees (Original Contribution by Fred Stephen, Department of Entomology, University of Arkansas and Timothy D. Paine, Department of Entomology, University of California, Riverside)

Pines defend themselves against bark beetles and fungal pathogens by both a preformed resin system and an induced hypersensitive response. In the latter, the infected area at the site of beetle/fungal attack initially becomes soaked with resin. A layer of cambium-derived periderm, which isolates the infecting organisms, subsequently develops around the resinous tissue (Berryman 1972; Lorio 1988, 1993; Paine et al. 1997; Stephen and Paine 1985; Wood 1972). Resins are typically mixtures of terpenes.

This field and laboratory experiment measures both the preformed and induced types of tree defenses and examines tree responses using trees of different age and growth rate as an indicator of vigor. Trees (for example Loblolly pine, *Pinus taeda* L., and shortleaf pine, *Pinus echinata* Mill.) will be inoculated with a blue-stain fungus, *Ophiostoma minus*, that is normally associated with the southern pine beetle (*Dendroctonus frontalis*), a serious tree killing insect in the South of the United States. Inoculation of this pathogen will result in the host tree producing an induced hypersensitive lesion.

Circular disks (1 cm diameter) of agar that were infected with the blue-stain fungus will be inoculated against the fresh phloem and sapwood of pines in a plantation. A similar-sized sterile wound will be made as a control. After 3 weeks the length of the hypersensitive and sterile lesions will be measured on all trees. Resin flow rate (vol/day) will be measured on each tree by scoring through the bark to the xylem with a 2.5 cm. diameter arch punch, and collecting the resin flowing from the wound during the subsequent 24 h.

Wounds and insertion of vials for resin collection, plus fungal inoculations will be made in the initial laboratory period. Resin collection must be made 24 h later. Three weeks later a subsequent lab will be devoted to exposing and measuring wound response lesions resulting from inoculations.

Each student group will inoculate six trees – three of small diameter and three of large diameter. Each tree will have four inoculations: two sterile and two fungal. Each tree will also have a core taken to determine radial growth and each tree will have two wounds to measure resin flow volume.

References

Berryman AA (1972) Resistance of conifers to invasion by bark beetle–fungus associations. Bioscience 22:598–602

Lorio PL Jr (1993) Environmental stress and whole-tree physiology. In: Schowalter TD, Filip GM (eds) Beetle–pathogen interactions in conifer forests. Academic, London, pp 82–101

Lorio PL Jr (1988) Growth and differentiation-balance relationships in pines affect their resistance to bark beetles (Coleoptera: Scolytidae). In: Mattson WJ, Levieux J, Bernard-Dagan C (eds) Mechanisms of woody plant defenses against insects: Search for pattern. Springer, New York, NY, pp 73–92

Paine TD, Raffa KF, Harrington TC (1997) Interactions among scolytid bark beetles, their associated fungi, and live host conifers. Ann Rev Entomol 42:179–206

Stephen FM, Paine TD (1985) Seasonal patterns of host tree resistance to fungal associates of the southern pine beetle. Z Ang Ent 99:113–122

Wood DL (1972) Selection and colonization of ponderosa pine by bark beetles. In: Van Emden HF (ed) R. E. S. Symposium No. 6. insect/plant relationships. Blackwell, Oxford

Interspecific Stimuli: Kairomones (Compounds Used for the Benefit of the Receiver); Responses to Prey Chemicals

Response of Hydra to Chemical Stimuli from Prey

Automatic discharge of *Hydra's* nematocysts releases glutathione from prey. Glutathione triggers contraction of longitudinal muscles of the *Hydra's* body and tentacles, and relaxation of ring muscles around the mouth (Hainsworth 1967).

1. Observe behavior (and record its sequences) of *Hydra* toward *Daphnia*
2. Record *Hydra's* response to different concentrations between 0.01% and 0.1% of glutathione

References

Hainsworth MD (1967) Experiments with Hydra and Sea-Anemones. In: Experiments in animal behaviour. Macmillan, New York, NY, p 29

Effect of Chemical Stimulation of Cladocerans, or Mosquito or Gnat Larvae

In a small transparent container with *Daphnia* that are evenly distributed in space, place a drop of meat extract with a hypodermic needle. Repeat with syrup. Observe the movement and clustering of the *Daphnia*.

For a different reaction, use a drop of a noxious stimulus such as ammonium hydroxide, potassium permanganate, or chloroform. Test dilutions in steps of 1/10, 1/100, 1/1,000, etc. to determine the sensitivity of the animals (Hainsworth 1967).

References

Hainsworth MD (1967) Behavior of Arthropods other than Insects. In: Experiments in animal behaviour. Macmillan, New York, NY, p 101

Chemical Attraction of Leeches

Leeches respond to chemical cues from potential hosts. Even a human fingerprint on the inner wall of a glass container triggers searching movements in a leech.

Feeding stimuli cause the medicinal leech, *Hirudo medicinalis*, to react with the sequence of probing, attachment, biting, and ingestion. A combination of NaCl and L-arginine triggers the complete sequence. The response is rather specific: D-arginine is not effective, and the analogs homoarginine and canavarine are active. Canavarine possesses all the three functional groups of arginine unchanged (Elliott 1986).

Among freshwater leeches, the Old-World *Erpobdella octoculata* feeds on prey such as *Tubifex* spp., *Chironomus* spp., and *Asellus aquaticus*. Living and freshly killed larvae of *Chironomus* sp., *Tubifex* sp., and *A. aquaticus* attract these leeches. Amino acids such as histidine and glutamic acid are the active stimuli (Kreuter et al. 2008).

In this experiment, you can test different water-borne chemical stimuli. Use water that had contained *Tubifex*, *Chironomus* larvae, or *A. aquaticus*; extracts of macerated prey animals worked into Agar; and the amino acids histidine and glutamic acid, alone or in mixture at concentrations above 5 mg mL^{-1}.

References

Elliott EJ (1986) Chemosensory stimuli in feeding behavior of the leech *Hirudo medicinalis*. J Comp Physiol A: Neuroethol Sensory Neural Behav Physiol 159:391–401
Kreuter K, Baier B, Aßmann C, Steidle JLM (2008) Prey location and prey choice by the freshwater leech *Erpobdella octoculata* using foraging kairomones. Freshwater Biol 53:1524–1530

Salamander Responses to Prey Extracts

Among the salamander species, good examples for feeding responses are the tiger salamander, *Ambystoma tigrinum*, a feeding generalist (Lindquist and Bachman 1980), and larval coastal giant salamanders, *Dicamptodon tenebrosus* (Chases 2008). Cover minced earthworm, or any meat, in a bag. Alternatively, prepare extracts of food items and work them into agar, then cut into cubes as food morsels. Offer to tiger salamander in a choice test. Describe the response. The full feeding sequence consists of orientation, approach, olfactory test, fixation, and snapping.

References

Chases LG (2008) The behavioral response of larval coastal giant salamanders, *Dicamptodon tenebrosus*, to chemical stimuli. MA Thesis. Humboldt State University, Biological Sciences
Lindquist SB, Bachman MD (1980) Feeding behavior of the tiger salamander. Herpetologia 36:144–158

Innate and Acquired Food Odor Preferences in Garter Snakes

Garter snakes respond to prey extracts with tongue flicking and attempts to bite (Gove and Burkhardt 1975). Test extracts of different prey animals such as earthworms, slugs, crickets, or leeches. Dip cotton balls or Q-tips into extracts and present to captive snakes. Experimenter should be behind screen to minimize snake's responses to light or movement.

References

Gove D, Burkhardt GM (1975) Responses of ecologically dissimilar populations of the water snake *Natrix s. sipedon* to chemical cues from prey. J Chem Ecol 1:25–40

Fish Feeding Responses to Amino Acids at Different pH Levels

Fish use amino acids in the water for both feeding and predator avoidance. Acidification of their water can change their feeding behavior. For example, Atlantic salmon, *Salmo salar*, normally are attracted to glycine, but avoid L-alanine. If the pH of their water is lowered from 7.6 to 5.1, the fish become indifferent to glycine, but are now attracted to alanine (Royce-Malmgren and Watson 1987). Attempt some version of an experiment with fish responses to amino acids at different pH levels.

References

Royce-Malmgren CH, Watson WH III (1987) Modification of olfactory-related behavior in juvenile Atlantic salmon by changes in pH. J Chem Ecol 13:533–546

Colored "Pastry" to Study Feeding Behavior in Birds

To present birds with "prey" that is differently flavored, a standard "pastry" can be prepared from flour and lard. A 3:1 mixture of flour and lard is mixed. To color the "food" lightly, 25-ml dye is added to 600 g pastry. For a darker color, use 70-ml dye for 600 g Pastry. Roll into a cylinder of 3 mm diameter. Cut into 1-cm sections (Edmunds and Dewhirst 1994).

References

Edmunds M, Dewhirst RA (1994) The survival value of countershading with wild birds as predators. Biol J Linn Soc 51:447–452

Detection of Buried Food (Seeds) by Birds

This experiment is best done in the laboratory or a zoo, rather than in the field.

Bury little bags with 1 g of seeds in soil of different depths. Use three treatments: unscented seeds, seeds scented with safflower oil, and seeds with pine oil. Give small rodents such as mice, gerbils, or hamsters, a choice of scented and unscented batches. Is there evidence that they detect buried sees by smell? Does the treatment affect the success rate in finding the hidden food? At what soil depth do the animals detect the food?

Repeat the same experiment with dry soil vs. moistened soil. Water content of soil is the most important factor for volatilization of organic compounds. In very wet soil, organic compounds desorb from soil (Vander Wall 2003).

Does the success rate in finding the seeds differ between the two treatments "wet" and "dry" soil?

References

Vander Wall SB (2003) How rodents smell buried seeds: A model based on the behavior of pesticides in soil. J Mammal 84:1089–1099

Interspecific Stimuli: Kairomones; Prey Responses to Predator Chemicals

Antipredator Responses by Intertidal Gastropods to Chemicals from Starfish

Sandflat snails (*Nassarius* sp.) leap away from starfish, their predator, and scallops (*Pecten* sp.) swim away in a peculiar manner from secretions of starfish.

Limpets (*Acmaea* sp.) raise their shell, rock or wiggle, and then move away at their highest possible speed (1 cm in 5 s) when a tube foot of a starfish contacts them. If the tube feet have succeeded in a tight grip, the limpet twists 90–180° around its vertical axis several times, breaking the grip by the starfish, and then moves away. The abalone, *Haliotis* sp., shows this response particularly well (Bullock 1953).

Extracts of tube feet or stomach juice from a starfish release these responses. Extract from a small starfish (*Leptasterias* sp., 2–3 cm from arm tip to arm tip) triggers escape responses in abalone which is 10 cm or longer!

Starfish hunt primarily in the low- and mid-tide zones. As a rule, high-tide species of gastropods and those living on brown sea weed do not show these responses (Bullock 1953).

References

Bullock TH (1953) Predator recognition and escape responses of some intertidal gastropods on presence of starfish. Behaviour 5:130–140

Response of Pond Snails to Chemicals of Leeches, Their Predator

1. Pulmonates such as *Limnea* sp., *Physa* sp., or *Bithynia* sp. retreat abruptly when encountering certain leeches (Kelly and Cory 1987; Rundle and Brönmark 2001).
2. For a baseline, record encounters, shell-twirling, and habituation.
3. Then brush a leech and use the contaminated brush to stimulate two snails.
4. Drip 0.1 ml leech water onto snail and observe responses.

References

Kelly PM, Cory JS (1987) Operculum closing as a defense against predatory leeches in four British freshwater prosobranch snails. Hydrobiologia 144:121–124
Rundle SD, Brönmark C (2001) Inter- and intra specific trait compensation of defence mechanisms in freshwater snails. Proc Biol Sci 268:1463–1468

Responses of Amphibian Larvae to Predator Cues

Larvae of amphibions flee and hide in response to waterborne chemicals from predators (Kats et al. 1988; Kats 1988; Petranka et al. 1987). Run water through a flow-through system of four consecutive tanks: from a water tank to a tank with predator fish, to a tank with tadpoles, and finally to a water tank. For the control experiment, there are no fish in the second tank. Observe and measure the spatial distribution and the shelter-seeking by the tadpoles under three conditions: in the second tank is no fish; a predatory fish; and nonpredatory fish.

References

Kats LB, Petranka JW, Sih A (1988) Antipredator defenses and the persistence of amphibian larvae. Ecology 69:1865–1870
Kats LB (1988) The detection of certain predators via olfaction by small-mouthed salamander larvae, *Ambystoma texanum*. Behav Neural Biol 50:126–131
Petranka JW, Kats LB, Sih A(1987) Predator-prey interactions among fish and larval amphibians: use of chemical cues to detect predatory fish. Anim Behav 35:420–425

Allelopathy

Bioassay of Juglone: Effect on Alfalfa Germination

Juglone (5-hydroxy-1,4-naphthoquinone) is a water-soluble yellow pigment that is exuded from leaves, fruits, bark, and roots of black walnut (*Juglans nigra*), English walnut (*J. regia*), and hickory (*Carya ovata*) trees, among others.

Juglone is responsible for the brown stain on hands when handling walnut husks. Seed husks and roots contain the highest concentrations of juglone. It has been extracted from the plant materials with a variety of solvents such as chloroform, methanol, or *n*-hexane. The plant contains the glycoside of α-hydrojuglone. This prevents self-poisoning. Upon damage to the plant tissue, it is hydrolyzed to glucose and α-hydrojuglone, and the latter then oxidized to juglone.

It is well known that vegetables such as tomatoes or lettuce do not thrive near walnut trees. Juglone and a few related compounds are responsible for this effect. At a concentration of 0.002%, juglone completely prevents germination of lettuce seeds. It is thought that rain wash from walnut leaves and exudation from roots transport juglone to the soil (Soderquist 1979; Terzi et al. 2003; Terzi 2008).

In this experiment, different concentrations of juglone (or extract from black walnut roots) are mixed into a medium. My colleague Dr. José Giner has used water-soaked cotton in a parafilm-sealed test tube. Alfalfa or lettuce seedlings in this medium are then measured for their growth. Measure radicle growth and seedling elongation separately. Compare germination of different species of plants in medium with and without juglon (for example, beans are not affected).

References

Soderquist CJ (1979) Juglone and allelopathy. J Chem Educ 50:782–783

Terzi I, Kocaçalişkan I, Benlioğlu, Solak K (2003) Effects of juglone on growth of cucumber seedlings with respect to physiological and anatomical parameters. Acta Physiologiae Plantarum 25:353–356

Terzi I (2008) Allelopathic effects of juglone and decomposed walnut leaf juice on musk melon and cucumber seed germination and seedling growth. Afr J Biotechnol 7:1870–1874

Environmental Odors

Choice of Pond Odors by Newts, Frogs, and Toads

Amphibians have been shown to discriminate the odor of their home pond from other mud or water odors (Forester and Wisnieski 1991; Grubb 1973, 1976; McGregor and Teska 1989; Ogurtsov and Bastakov 2001). In a two-way choice apparatus, an animal chooses between mud or water from the home pond and some control mud or water from a different source. This experiment tests the ability to home by chemical cues, one aspect of orientation in space use newts, frogs or toads, depending on animals available.

References

Forester DC, Wisnieski A (1991) The significance of airborne olfactory cues to the recognition of home area by the dart-poison frog *Dendrobates pumilio*. J Herpetol 25:502–504

Grubb JC (1973) Olfactory orientation in *Bufo woodhousei fowleri*, *Pseudacris clarki* and *P. streckeri*. Anim Behav 21:726–732

Grubb JC (1976) Maze orientation by Mexican toads, *Bufo valliceps*, using olfactory and configurational cues. J Herpetol 10:97–104

McGregor JH, Teska WR (1989) Olfaction as an orientation mechanism in migrating *Ambystoma maculatum*. Copeia 1989:779–781

Ogurtsov SV, Bastakov VA (2001) Imprinting on native pond odour in the pool frog *Rana lessonae* CAM. In: Marchlewska-Koj A, Lepri JJ, Műller-Schwarze D (eds) Chemical signals in vertebrates, vol. 9. Kluwer, New York, NY, pp 433–43825   Further Possible Experiments

Index

A

Abalone, *Haliotis sp.*, 144
Abdominal glands, moths, 135
Acer rubrum, 77
Acer saccharum, 77
Acetates, 135
Acidification of water, 143
Acorn parts, 83
Acorns, 33, 84
Acyanogenic plants, 73
Adsorption, 65, 76
Adult-type growth stages of trees, 44, 94
Adventitious shoots, 94
Agar gel, 82, 142
Agarose, 83
Age of scent trail, 136
Aggressive dominance, 116
Alarm secretion, 134
Alfalfa germination, 146
Alkaloids, 60, 94, 104, 108
Allelopathy, 146
Allomones, 139
Allopatric carnivores, 20, 28
Amazon basin, 76
Amazon forest, 65
American basswood (*Tilia americana*),
     77, 94
American beech (*Fagus grandifolia*), 77, 79
Amino acids, 8, 131, 142, 143
Ammonium hydroxide, 141
Amphibian larvae, antipredator behavior, 145
Androgenized females, 112
Anhydrous ferric chloride, 78
Animal behavior, 112
Anogenital distance, mice, 117
Anti-microbial functions, 76
Antipredator behavior, amphibian larvae,
     134, 145
Ants, 3, 65, 136

Aphids, 3
  honeydew, 4
  myrmecophilous, 4
  attended aphids, 4
  Pea aphids, 4
Apocrine gland, 122
Apodemus sp, 20
Apple seeds, 71
Apple trees, 38, 71, 79, 136
Arabidopsis thaliana, 70
Arboreal predators, 27
Argentine ant, 136
Asellus aquaticus, 142
Ash (*Fraxinus spp.*), 80
Aspen (*Populus spp.*), 44, 48, 50, 82, 94
Atlantic salmon, *Salmo salar*, 143
Attractiveness rating, humans,135
Avigon, 14
Avoidance of predator odors, 25
Axillary glands, 122
Axillary odor, 122

B

Badger, 51
Bait
     birds, 14–16
     cottontails, 37–40
     deer mice, 131
     fish, 7–11
     squirrels, 25, 33, 60
Ballpoint pen ink, 136
Bamboo, 70
Bark beetles, 139
Bark, 83
Basswood (*Tilia sp.*), 44, 80
Bat(s), 64
Beans, 146
Bears, 64

Beaver (*Castor canadensis*) 20, 46, 53, 64, 76, 77
Beaver castor, 55
Beaver habitat, 54
Beech (*Fagus spp.*), 77, 79
Beet armyworm (*Spodoptera exigua*), 99
Belching, 128
Bioassay, 108, 112
    of tannins, 108
Bird Shield, 14
Birds, 14, 32, 69, 76
    buried seeds detection, 144
Birdsfoot trefoil (*Lotus corniculatus*), 71
Bithynia sp. (freshwater snail), 145
Black locust (*Robinia pseudacacia*), 46
Black walnut (*Juglans nigra*), 146
Blackwater rivers, 76
Blue-stain fungus, *Ophiostoma minus*, 140
Body odors, 122
Bovids, 44
Bovine serum albumin (BSA), 82
Bud scales, 83
Bullhead, brown (*Ictalurus nebulosus*), 11
Bullhead, yellow (*Ictalurus natalis*), 11
Buried food detection, 144
Burping, 127, 129
Burying of seeds, 33
Butternut (*Juglans cinerea*), 27

C
Cache, food, 32
Caching, 32
Cafeteria-style food choice experiment, 44
Caffeine, 131
Cambium, 140
Camera trap, 56
Camphor, 131
Canada geese, 14
Canada mayflower (*Maianthemum canadense*), 64
Canals (beaver), 54
Canavarine, 142
Capsaicin, 61, 131
Capsicum sp., 60
Carcasses, 65
Carnivorous fish, 8
Cascade of C_{18} compounds, 99
Cassava (*Manihot esculenta*), 70
Castor canadensis (North American beaver), 19
Castoreum, 55
Cat repellent, 131
Cat urine, 131
Caterpillars, 105

Cats, 20, 27, 53, 131
Cellulose membrane filters, 108
Chancho (hoatzin), 65
Chaoborus (phantom midge), 85
Chemical alarm responses, 131
Chemical attractants, 135
Chemical attraction, 141
Chemical defense, 44, 64
    insects, 139
Chemical gradient, 32
Chemical hunting, 8
Chemical lures, 8
Chemical trails, 136
Chemically induced defenses, 85
Chipmunks, 21, 64
Chironomus spp., 142
Chloroform, 141, 146
Choice experiment, 108
Chrysomelidae, 70
Cladoceran *Bosmina longirostris*, 86
Cladocerans, 141
Clark's nuthatches, 33
Clay, 65
Coastal giant salamander, *Dicamptodon tenebrosus*, 142
Coelomic fluid, 134
Coffee, 128
Commercial insect chow, 104
Common shiner (*Notropis cornutus*), 131
Competitiveness, 118
Competitors, 32
Concentration effects, 104, 108, 133
Concord grapes, 14
Coniferous trees, 76, 77
Conjugation, 76, 125
Constitutive chemical defenses, 94
Copepod *Eudiaptomus gracilis*, 86
Copper II ethylacetoacetate, 70
Cottontails, *Sylvilagus floridanus*, 37, 39, 76, 77
Cottonwood, Eastern (*Populus deltoides*), 44, 48, 96
Cotyledons, 32
Counter-marking, 112, 116
Cow, 27
    manure, 22
Coyotes, 20, 26, 60
Crataegus sp, 77
p-cresol, 76
Crickets, 143
Cucumbers, 128
Cut-off branches, 54
Cyanogenesis, 72, 73
Cyanogenic compounds, 69
Cyanogenic glucoside, 70
Cyanogenic glycosides, 70

Cyanogenic plants, 70, 72
Cytochrome P450, 128

D

Dams (beaver), 54
Daphnia (Cladocera, Crustacea),
    86, 87, 141
D-arginine, 142
Deciduous trees, 77, 79, 83
Deer, 26, 32, 38, 44, 60, 64, 76, 77, 104
Deer mice, 33, 131
Defense compounds, plants, 43
Defense secretions, animals, 139
Demes, 112
Desorption from soil, 144
Deterrency Index, 105, 109
Detoxication, 64
Detoxication mechanisms, 128
Dhurrin, 70
Dimethyl anthranilate, 14
Diplocardia riparia, 134
Discrimination, body odors, 121
Dithiolanes, 20
Dogs, 53, 131
Domestication, 70
Dominant males, 112, 116
Dorsal pores, earthworms, 134
Double blind test, 104
Douglas fir, 79

E

Earthworms, 134, 143
    alarmed, 134
Eastern cottonwood, (*Populus deltoides*), 93
Eastern tent caterpillars, 135
Eisenia foetida, 134
Electivity Index, 44, 46
Embryo (seed), 32
English walnut (*Juglans regia*), 146
Environmental odors, 147
Epicotyl, 32
Erpobdella octoculata, 142
Eructation, 144, 145
Euphorbia lathyris, 39
European buckthorn, 79
Excretion(s), 64, 128
Extrafloral nectaries, 65

F

Fagus grandifolia (American beech), 77
False hellebore (*Veratrum viride*), 39
(E)-β-Farnesene, 4

Feeding, 143
Feeding bed, 46
Feeding behavior, birds, 143
Feeding ecology, 131
Feeding inhibition, 108
Feeding inhibitors, 76
Feeding repellent, 13, 59
Feeding responses, 143
Feeding stimuli, 108, 142
Feigl–Anger test, 70
Female-female communication, 112, 116
Feral dogs, 20
Fern, 64, 99
Ferric chloride ($FeCl_3$), 78
Field grid, 37
Fish, 8, 145
Fisher (*Martes pennanti*), 64
Flea beetle *Phyllotreta nemorum*, 70
Food choice experiments, 38
Food choices, 43, 60, 131
Food odor preferences, 143
Food plants, 131
Food preferences, insects, 108
Food processing, 33, 76
Forest, chemical ecology in, 63
Fox urine, 131
Foxes, 20, 26, 60
Fox, red (*Vulpes vulpes*), 26
Fragrance, 131
Free-ranging mammals, 51
Freshwater leeches, 142
Frog tadpoles, 131
Frogs, 147
Fungal pathogens, 139
Fungi, 60

G

Gallic acid, 48, 108
Galloyl glucoses, 108
Garlic, 128
Garter snakes, 128, 143
Gender discrimination, 122
Gene activation, 100
Generalist herbivore, 44
Generalists, 104, 108
Genetic dispositions, 112
Geometric isomers, 135
Gerbils, 137, 144
Germination, 32
Giant river otter (*Pteronura brasiliensis*), 65
Glucuronic acid, 128
Glutamic acid, 142
Glutaraldehyde, 87
Glutathione, 141

Glycine, 143
Glycoside of α-hydrojuglone, 146
Gnat larvae, 141
Goat, 27
Golden bamboo lemur (*Hapalemur aureus*), 70
Goldenrod (*Solidago sp.*), 5
Goldfish urine, 136
Grackles, 32
Grapes, 82
Gray fox, 20
Gray squirrel (*Sciurus carolinensis*), 26, 64, 131
Grazers, 85
Great horned owl (*Bubo virginianus*), 20
Green alder (*Alnus crispa*), 94
Grid, experimental, 72
Grouse, 131
Guignard sodium picrate test, 71
Gustatory responses, 131

H
Hamamelis virginiana (witch hazel), 48, 77
Hamsters, 137, 144
Hawks, 20
Hawthorn (*Crataegus spp.*), 77
Hemlock (*Tsuga sp.*), 79
Hepatic microsomal
        cytochrome P450, 128
Herbal teas, 128
Herbivore repellent, 38
Herbivore urine, 28
Herbivores, 38, 39, 43, 60, 70
Herbivorous insects, 76, 104, 108
Herbivorous vertebrates, 76
Herbivory, 75, 77, 93–98
Herring gulls, 14
(*Z*)-9-Hexadecanal, 136
Hickory (*Carya ovata*), 146
Hickory nuts, 27
Histidine, 142
Hoatzin (*Opisthocomus hoazin*), 65
Home pond water, 147
Home ranges, 112
Homoarginine, 142
Hormonal influences, 137
Hormonal status, 112
Hornworm caterpillars, 104, 108
Hot peppers, 128
House mice (*Mus musculus*
        or *M. domesticus*, 112
Howler monkeys, 65
Human behavior, 122
Human body odor, 121
Human fingerprint, 141

Human scent, 122
Hydra sp., 141
Hydrogen cyanide (HCN), 70
α-Hydrojuglone, 146
Hypocotyl, 32

I
Induced chemical defenses, 94
Induced defense, 93
Induced hypersensitive lesion, 140
Induced hypersensitive response, 139
Inducible defense, 85
Info-chemicals, 86
Insect feeding behavior, 103, 107
Insect herbivory, 95
Insects, 70, 76, 82, 139
Interspecific behaviors, 4, 139
Interspecific stimuli, 141, 144
Intertidal gastropods, 144
Intraspecific communication, 133
Intrauterine hormonal stimulation, 112
Isosulfan blue, 136
in utero hormonal effects, 117

J
Jasmonate, 99
Jasmonic acid, 99
Jays, 32 Juglone, 146
Juglone (5-hydroxy-1,4-naphthoquinone), 146
Jungle fowl, 14
Juvenile-type growth form,
        trees, 44, 94

K
Kairomones, 141, 144
King vulture, 65

L
Ladybird beetles (coccinellids), 4, 139
Lake sturgeon (*Acipenser fulvescens*), 131
Lake, 8
L-alanine, 143
Larch (*Larix spp.*), 78, 79
L-arginine, 142
Larix sp., 78
Latency, 113
Latrine sites, 53
Leaching unpalatable compounds, 76
Leaf disk assay, 108
Leaf disk choice test, 104, 108

Leaf sections (leaf disks), 108
Leaf size, 94–95
Leaf-eating bird, 65
Leaves, tannin test, 82
Leech chemicals, 141
Leeches, 141, 143, 145
Legume species, 71
Lettuce, 146
Limnea sp, 145
Limpets (*Acmaea sp.*), 144
Linamarin, 70
Live traps, 20
Liver, 128
Livestock pastures, 131
Loblolly pine, *Pinus taeda*, 140
Lodge, beaver, 43, 44
Lotaustralin, 70
Lotus corniculatus (birdsfoot trefoil), 70–72
Lugol's solution, 87
Lumbricus terrestris, 134
Lye, 33

M
Macaws, 65
Madagascar, 70
Major Histocompatibility Complex, 112
Malus sylestris, 77
Mammal repellents, 60
Mammals, 60, 76
Manduca sexta, 108
Manihot esculenta, 70
Manioc, 70
Marking, 64
Marsupials, 44
Mastication, 128
Mate choice, 112, 137
Meat extract, 141
Medicinal leech (*Hirudo medicinalis*), 142
Metabolic deactivation, 128
Methanol, 146
Methyl anthranilate, 13, 60
Methyl jasmonate, 99
8-methyl-*N*-vanillyl-6-nonenamide, 60
Mice, 60, 64, 112, 115, 131, 144
Mice, marking territories, 112
Microorganisms, 70, 82
Microtus sp. (voles), 20
Middens, 64
Minnow traps, 7
Mixtures, 8
Molluscs, 70
Mosquito larvae, 141
Moths, 135

Mouse scent marks, 138
Mucus, 134
Mucus trails, 134
Mus domesticus, 112
Mus musculus, 112
Mushrooms, 64
Mustelid anal gland, 20
Mustelids, 20, 26
Mutualistic fungus, 99

N
NaCl, 142
Neckteeth, *Daphnia*, 85
Neighbor effect, 39
Nematocysts, 141
Nematodes, 133
Newts, 128, 147
n-hexane, 146
Ninhydrine, 138
Northeastern forest, 64
Northern red oak, 77
Norway rats, 131
Norway spruce, 78
Noxious stimulus, 141

O
Odor-carrying lipids, 138
Odors, 131
Oil of citronella, 131
Olfactory individual recognition, 122
Olfactory recognition, 122
Olfactory signals, 112
Olfactory gender recognition, 122
Open field test, 111, 115
Open-field arena, 116
Opisthocomus hoazin, 65
Orkney Islands, 20
Orkney voles (*Microtus arvalis orcadensis*), 20
Osmetrichia, 138
Ostrya virginiana, 77
Overshadowing, 39
Owls, 26
Oxidation, 76, 128

P
Populus grandidentata, 48
Populus tremula, 76
Populus tremuloides, 77
Parental territory, 112, 116
Pastry for birds, 143

Peeled sticks, 54
Pericarp, 32
Periderm, 140
Petioles, 32
pH levels, 143
Phenol, 76
Phenolate ion, 76
Phenolic compounds, 104, 108
Phenolics, 48, 75, 76, 94, 104, 108, 131
Phenolics test, 98
Pheromone
    alarm pheromone, 4
    signaling pheromone, 112
Pheromone plume, 135
Pheromone responses, 135
Pheromone traps, 135
Pheromones, 133
    priming pheromone, 112
Phloem, 140
Phytoplankton, 85–90
Picea abies, 78
Pigeons, 14
Pine oil, 144
Pine trees, 139
Pinus strobus (white pine), 78
Pinus sylvestris (Scots pine), 78
Plant chemistry, 104
Plant defense, 139
Plant secondary compounds. *See* Plant
    secondary metabolites
Plant secondary metabolites (PSMs), 44, 76,
    94, 128, 133
Plant volatiles, 99–100, 128
Plants, cyanogenic, 69
Plethodon, 137
Poisonous weeds, 131
Polymorphism, 72, 73
Polyphenols, 82
Pond, 8, 55
Pond odors, 147
Pond snails, 145
Poplar (*Populus spp.*), 76
Population density, 55
Populus, 93
Populus tremuloides, 77
Porcupines, 76, 77
Potassium permanganate, 141
Potato plants, 104, 108
Predator avoidance, 143
Predator chemicals, responses to, 144
Predator fish, 145
Predator odors, 19, 25
Predators, 85
Predator "sign", 26

Predatory fish, 7, 131
Predatory snake odors, 131
Preformed resin system, 139
Prenatal effects, 118
Prey chemicals, 141
Prey extracts, 131, 142
Prey fish, 131
Prey odors, 7, 131
Proteins, 104, 108
Pseudobush (pseudoshrub), 39, 47
Pulmonates, 145
Pumpkinseed, 131
Pyrocatechol, 76
Pyrogallol, 76

Q
Quaking aspen (*Populus tremuloides*), 48, 80, 97
Quercus rubra (red oak), 97

R
Rabbit repellents, 38
Rabbit, (*Oryctolagus cuniculus*), 39, 40, 131
Raccoons, 26, 60
Radial diffusion assay, 33, 81
Radicle, 32
Radicle growth, 146
Rats, 131
Recruitment trails, 135
Red foxes, 20
Red maple (*Acer rubrum*) 48, 64, 77, 79
Red oak (*Quercus rubra*), 45, 107
Red squirrel (*Tamiasciurus hudsonicus*), 64
Red wine, 82
Red-spotted newts (*Notophthalmus
    viridescens*), 131
Reduction, 125
Red-winged blackbird
    (*Agelaius phoeniceus*), 14
ReJeX-iT, 14
Repellents, 76, 131
Resident breeding females, 116
Resin defense, 139
Resin flow rate, 140
Rhamnus cathartica, 78
Ring-necked pheasants, 14
Rock bass (*Amblopites rupestris*), 11
Rodents, 20, 44, 60, 64, 131
Rosaceae, 136
Rotifer *Brachionus calyciflorus*, 87
Rotifers, 86
Ruffed grouse
    (*Bonasa umbellus*), 32, 76

S

Safflower oil, 144
Salamander, 137, 142
Salicin (saligenin glycoside), 76
Salicortin, 76
Saliva of caterpillars, 100
Salix sp., 78, 93
Sandflat snails (*Nassarius sp.*), 144
Sapwood, 140
Scallops (*Pecten sp.*), 144
Scats, 64
Scavengers, 65
Scenedesmus (Chlorococcales, Chlorophyta), 86
Scent communication
    intersexual, 112
Scent marking, 53, 137
    absence,118
    frequency, 137
    mammals, 112
    mice, 115
Scent mounds, 54, 56
Sciomyzid fly larvae, 134
Scots pine, 78
Secondary plant metabolites (compounds).
    See Plant secondary metabolites
Secretions, marking with, 64
Seedling elongation, 146
Selection, plants by herbivores, 70
Sense of smell, 32, 65
Sex differences, 117
Sheep, 27, 138
Shelter-seeking, tadpoles, 145
Shortleaf pine, *Pinus echinata*, 140
Short-tailed shrews, 21
Shrews, 20
Skin gland size, 137
Skin glands, 122
Slugs, 64, 134, 143
Small mammals, 20, 26, 131, 138
Small rodents, 144
Smallmouth bass (*Micropterus dolomieui*), 131
Snakes, 131, 134
Snowshoe hare (*Lepus americanus*), 64, 93
Soaking, twigs, 33
Social odors, mice, 111
Sodium lauryl sulfate, 138
Soil water content, 144
Soiled arena, 118
Solanaceae, 104, 108
Solvents, 146
Sorghum bicolor, 70
Southern pine beetle (*Dendroctonus frontalis*), 140
Southern United States, 76

Soybean cyst nematode, *Heterodera glycines*, 133
Spawning pheromone, 136
Specialists, 104, 108
Species specificity, 133, 135
Spicy food, 125, 126
Spines, rotifers, 86
Spurges (*Euphorbia spp.*), 39
Squirrel repellents, 60, 131
Squirrels, 25, 31–34, 60
St. Johnswort, 71
Starfish (*Leptasterias sp.*), 144
Starfish chemicals, 144
Starlings, 14
Steeplebush (*Spiraea sp.*), 5
Stick insects, 139
Stimulants, 135
Stinkbugs, 139
Storage in soil, 33
Stream, 8
Striped maple (*Acer pensylvanicum*), 45
Subcaudal gland, 53
Sugar maple (*Acer saccharum*), 64, 77
Sugars, 104, 108
Suint, sheep, 138
Sulfides, 20
Sulfur compounds, 20, 28
Surfactant, 138
Sweat gland, 122
Sympatric predators, 20
Syracuse, New York, 70

T

Tadpoles, 145
    behavior, 131
Tall fescue (*Lolium arundinaceum*), 100
Tangarana tree, 65
Tannic acid, 82, 104, 108
Tannin content, 32
Tannins, 31, 82, 104, 108
Tapirs, 65
Taproot, 32
Tea, 82, 131
Tent caterpillars, 135
Termites, 136
Terpenes, 100, 140
Terpenoids, 93, 104, 108
Territory, 53, 116
Test for total phenolics, 75
Testosterone, 137
Tetrodotoxin, 125
Thietanes, 20
Thiols, 20

Tiger salamander, *Ambystoma tigrinum*, 142
Tilia americana, 77
Tilia sp., 93
T-maze, 112–114
Toads, 147
Tobacco hornworm, *Manduca sexta*, 99, 104
Tolerance, 125
Toluene, 72
Tomato, 104, 108, 146
Tongue flicking behavior, snakes, 131, 143
Torture tree, T*riplaris sp*, 65
Toxins, 65
Tracks, 54
Trail camera, 47
Trail following, 134–135
Trail pheromone, 136
Trail pheromones, ants, 136
Trails, 55
Transect, 72
Translocation, 76, 77
Trapping success, 131
Tree stumps, 54
Trees, 77, 79
Tremulacin, 76
Tremulone, 76
Trifolium repens (white clover), 72
Trigeminal nerve, 60
Tripartite regulatory mechanism, 100
T-shirt experiment, 121
Tubifex spp., 142
Turkeys, 32
Two-way choice apparatus, 107, 112, 147
Two-way choice test, 111, 112

U
Underground storage, seeds, 32
Urine, 76
    mark(ing), 112, 116, 117, 138

V
Vanillic acid, 133
Vanilloid receptor subtype 1, 60
Vegetation, 54
Ventral glands, 137

Vertebrate herbivores, 70, 82
Visualization, 138
    pheromone pulses, 136
Volatilization of organic compounds, 144
Voles, 60, 76, 77
Volicitin [n-(17-hydroxylinolenoyl)-l-
    glutamine], 100

W
Walnut trees, 146
Water flea, 87
Water-borne female odors, 131
White clover (*Trifolium repens*), 70
White oak, (*Quercus alba*), 32
White pine (*Pinus strobus*), 78
Whitewater rivers, 76
Wild raisin (*Viburnum sp*), 5
Wildlife, 76
Willow (*Salix spp.*), 44, 76, 78, 79
Wind tunnel, 135
Witch hazel (*Hamamelis virginiana*), 45, 46,
    48, 77, 79
Witch hobble (*Viburnum lantanoides*), 48
Wolves, 20, 26
Woodchucks, 60
Woodland jumping mice (*Napaeozapus
    insignis*), 19
Wool fat, 138

X
Xenobiotics, 125
Xylem, 140

Y
Yellow birch (*Betula alleghaniensis*), 45, 64
Yellow-headed vulture (*Cathartes sp.*), 65
Yellow perch (*Perca flavescens*), 11
Yuca (cassava), 70
Y-Maze, 112, 137

Z
Zooplankton, 86

Printed in the United States of America